No. 1602
$15.95

EXPERT TECHNIQUES FOR HOME VIDEO PRODUCTION

BY E. GRAYSON MATTINGLY

TAB BOOKS Inc.
BLUE RIDGE SUMMIT, PA. 17214

I would like to dedicate this
book to my wife, Suzanne, my
daughter, MaryBeth, and my mother
and father, Elsie and William.

FIRST EDITION

FIRST PRINTING

Printed in the United States of America

Library of Congress Cataloging in Publication Data

Mattingly, E. Grayson (Eugene Grayson), 1941-
Expert techniques for home video production.

Includes index.
1. Home video systems. 2. Television—Production and direction. I. Title.
TK9960.M28 1983 778.59'9 83-4946
ISBN 0-8306-0102-3
ISBN 0-8306-0602-5 (pbk.)

All photographs and drawings are by the author unless otherwise credited. All diagrams are by Suzanne Fisher Mattingly.

Contents

Introduction v

1 **The TV Process** 1
The Television Picture—Tubeless Cameras—Recording the Television Picture—Helical Scan Recording—Playback—Color and Television—Deriving the Color—Color Playback

2 **Controlling and Monitoring the TV Signal** 10
Impedance—Splitting a Signal—Stabilizing Signals—The Processing Amplifier—Time Base Corrector—Monitoring the TV Signal Using the TV Set—Using a Receiver/Monitor—Other Means of Monitoring the RF and Video Signal

3 **Sound Advice about Audio** 20
Selecting the Microphone—High-Level Signals—Microphone Mixers—Checklist for Recording Audio—Audio Dub—Blanketing Reverberation—Processing Sound—Planning Your Audio Production—Production Tips—Production Precautions

4 **Lighting Methods and Techniques** 39
Color Temperature—Light Intensity—Monitoring Light Intensity—Lighting the Environment—Direct Lighting Techniques—Indirect Lighting—Mood—Production Tips

5 **Electronic Editing** 52
Editing Setups—Editing Modes—Editing Strategies—No-Script Programs—Double Shoot—Scripting in the Field—Fully Scripted Programs—Editing on the Run—Live-Action Editing—Logging and Indexing the Tapes—Animation—Video Duplication—Production Tips—Editing Tips

6 Multicamera Systems 68

Multicamera Hookup—Matching Cameras—SEG Operation—Directing the Show—Script Directions—Putting It Together—Production Tip

7 Incorporating Other Media 79

Slide-to-Tape Transfer—Lecture and Slide Presentation—Dissolver—Hole in the Screen—Film-to-Tape Transfer—Titles—Artwork—Expanding Your System

8 Video Grammar 95

Camera Movements—Point of View—The Eyeline—Transitions—Audio Grammar

9 Preproduction Guidelines 108

10 Production Models 114

Video Fun and Games—Instant Replay—Role Play—Role Play Models—The Video Catalog—The Video Scrapbook—Special Event Video—The Training Tape—Video Star—Legal Video—Video Deposition—The Video Will—Day-in-the-Life-of Video—Environmental View—Bedroom Documentaries

Recommended Reading 137

Glossary 140

Index 168

Introduction

When video technology was young, it was fun, but it was wearing on the mind and back! A lot has happened in the last several years; the gear is lighter and a lot more sophisticated. I don't know how far we'll go with video. Technology has a way of making things obsolete. For now it seems that we are approaching the golden age of video, a period of using video in the home to entertain, to train, and to educate. It's a wonderful outlet for our creative energies. There's nothing like doing it yourself in living color.

Before we are carried away on the magic carpet of video, let's not forget that old saying, "You may have instant replay, just don't confuse it with instant success!" This book contains what I feel are real shortcuts to instant success in using your home video system.

In some instances, especially in the first two chapters, you might think you are headed for ED TECHNIVILLE! But have no fear; an engineering degree is not necessary for creating good video productions. On the other hand, an understanding of the TV process will certainly pay off in the long run. Just grit your teeth. After the first two chapters, you can shift into high gear!

Chapter 3 surveys the many types of microphones available in today's market. A discussion of the setup and operation of mixing gear is followed by a review of audio production techniques for video recording. The chapter on lighting methods and techniques presents a practical approach to lighting. It covers "theoretical" aspects of lighting for TV and gives many tips and guidelines for lighting indoor and outdoor taping setups.

Electronic editing is an exciting new aspect of home video production. Chapter 5 covers the details of editing equipment setups and operation. It also gives you a number of practical and effective editing production strategies.

Initially you will be concerned with the setup and operation of a single camera system, but at some point you might want to try using more than one camera at a time. Chapter 6 explains how to do it. The chapter on incorporation of other media describes low-cost, effective strategies for incorporating slides, photographs, films, home computers, and graphics into video productions. Chapter 8 will help you develop techniques and skills that will add continuity and interest to your productions. Camera moves and points of view are discussed, as well as audio and video transitions. This chapter will have you watching your favorite TV program from an entirely new perspective.

The key ingredient to a successful video production, regardless of its length or size, is effective planning. Chapter 9 outlines a tested and proven production planner that will save valuable time in the preparation of video productions.

The chapter on production models explores the many applications of video, from video depositions to bedroom documentaries. The specific production tips and guidelines accompanying each individual production model give practical, realistic directions for using video for fun or profit!

Whatever you plan to do with your video system, the information in this book will make it easier. And I hope that you will enjoy sharing your experiences and skills with others as much as I have.

Chapter 1

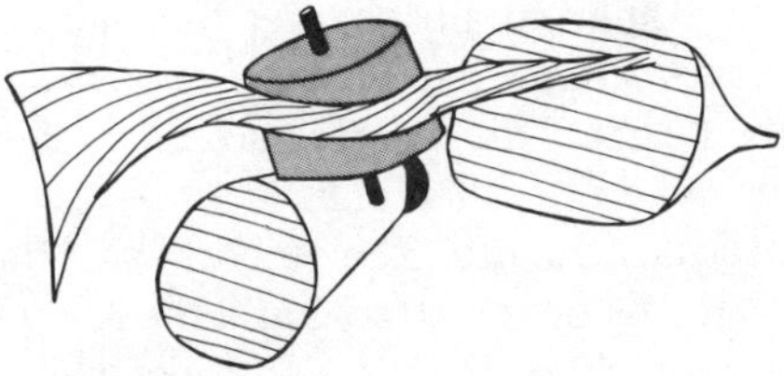

The TV Process

Television communication, like all other communication media, requires the proper management and application of certain tools. The basic tools used by a television communicator are: a camera, a videocassette recorder, a microphone, and a TV receiver. In addition to learning which buttons to push and which knobs to turn, one must also understand how these separate pieces of equipment fit together and the contribution each makes to the formation of electronic information.

Although a complete, but academic, understanding of all the established theories and proven techniques used in television communication will *not* guarantee the producer successful video productions, a practical understanding of the TV process will assure the creation of higher quality video programs. In addition, understanding the TV process should provide the video user with a cost-effective and innovative means of using today's equipment, and a commonsense approach to coping with the ever-changing "state-of-the art" tools of the television trade. I will begin with an explanation of how a black-and-white picture is formed, recorded, and played back.

THE TELEVISION PICTURE

In the television process, light reflected from a subject is focused by the lens on the photoelectrically sensitive inner face of the camera tube, or pick-up tube, located inside the camera. The electronic characteristics of the pick-up tube face cause it to respond differently to areas of light and dark in the focused image.

An electronic "gun" at the rear of the pick-up tube generates a steady stream of electrons that scans the rear side of the tube face in a very specific way. As the *steady stream of electrons* generated by the gun strikes the differentially charged tube face, it is proportionally changed into a *fluctuating stream of electrons* and is picked up by a plate called the target. This electron stream is then amplified and modulated to become the video signal.

But let's back up a minute. Remember, I said that the stream of electrons generated by the electron gun at the rear of the tube scans the tube face in a very specific way. Let's take a look at exactly how that happens.

In the system of television used in the United States, the scan pattern is limited to 525 horizontal lines, scanned left-to-right every 1/30 of a second. This means that every 1/30 of a second the camera "sees" one complete picture. But it is not really as simple as that. The electron beam performs *two* scans of pick-up tube in that 1/30 of a second. On the first trip, it scans the odd numbered lines; on the second trip it scans the even numbered lines. Each scanning trip takes 1/60 of a second and is referred to as one *field*.

Because our eyes tend to retain an afterimage of what we see for a brief period of time after the object has disappeared, the alternating odd and even scanning trips or fields appear to blend into one continuous picture that we call a *frame*. In order to make this process visually believable, the separate fields must be precisely interlaced—horizontally and vertically. This is the job of the synchronization signal or *sync signal*. Problems in this area will show up immediately. The sync signal, which may be generated either in the camera or by an external sync generator, serves the same function as the film advance mechanism in a motion picture camera. It orders precisely the succession of fields and frames and keeps them from overlapping or slipping horizontally. Synchronization is included in the video signal developed by the camera. We call such a signal *composite video*. Composite video carries both picture information and the order in which that information will be played back. But played back how?

The television picture is played back in just the reverse order in which it was originally developed. In the rear of the cathode ray tube, which forms the heart of the television receiver, is another electron gun. The incoming, composite video signal causes this gun to scan the photoelectrically sensitive layer of phosphorus on the inside of the cathode ray tube's display in a scanning pattern precisely like that performed by the electron gun in the camera's pick-up tube.

With a live transmission the process is instantaneous. The fluctuating beam striking the inside of the CRT's display face causes the phosphoric layer to emit light. The stronger the beam the more light is emitted. As the beam travels across and down the tube face, the dots of light make up a pattern that is similar to the one originally resulting from the image focused on the face of the camera's pick-up tube. As each field succeeds the other, we "see" one complete picture transmitted at the speed of light from the camera to the television face.

TUBELESS CAMERAS

Hitachi recently introduced a camera that uses a metal oxide semiconductor (MOS) device to pick up images rather than the conventional camera tube. The MOS is a solid-state component composed of hundreds of thousands of light receptor transistors mounted on a silicon chip. Light reflected and focused on the chip is scanned electronically and transformed into a fluctuating beam of electrons that forms a video signal. RCA has also introduced a solid-state pick-up device called a Changed Coupled Device or CCD. These pick-up chips have the following advantages over the conventional tubes:

- ☐ Longer life (10 to 20 times longer).
- ☐ No possibility of picture burn or lag.
- ☐ Lower power consumption.
- ☐ Lighter weight and smaller camera configuration.
- ☐ Instant warm up.

RECORDING THE TELEVISION PICTURE

To record the camera's picture for replay at another time or place, we break the link between the camera and the TV set with a device called a videocassette recorder. This device is often referred to by the acronym VCR. If the recorder accepts videotape stored on open reels, it can be called a video tape recorder or VTR. See Fig. 1-1.

In simple terms, the VCR takes the composite video signal and records it on magnetic tape in a manner similar to that in which audio signals are recorded. Nevertheless, there are some important differences between the two procedures. Most of the differences are due to the need to record a great deal more information to reproduce

Fig. 1-1. In the earlier open-reel VTRs (video tape recorders), the tape was wrapped around the head drum at a slant.

a video signal—nearly 100 times more! Obviously, running tape past stationary heads, as in audio recording, won't do. To store the amount of information required for a video picture, the tape would have to move at 55 feet per second. That's nearly 36 miles per hour! At that rate, a one-hour videotaped program would require approximately 198,000 feet of tape!

HELICAL SCAN RECORDING

There are other, more practical ways of recording video signals. The one you are most likely to encounter is called the *helical scan* recording. This process is common to all 1-inch., ¾-inch (or "U-Matic"), ½-inch (Beta or VHS), and ¼-inch video recorders—whether cassette, cartridge, or open reel. Helical scan recording came about when someone realized that a specific area of tape would hold a lot more information if the tracks were slanted. Using the slant-track, or the helical scan approach, you can compress information requiring 55 feet of tape in straight track recording into just 7½ inches of tape. Therefore, tape storage and handling is less of a problem.

Regardless of the format of your VCR, all helical-scan machines function basically the same. When a cassette is placed in your VCR and the play or play/record button is depressed, the tape is pulled out of the cassette and, first, passes over the bulk erase head. It is then wrapped around a drum assembly. The drum assembly is mounted on a slant. It has two video heads mounted on its side (180 degrees apart). The head drum rotates as the tape is pulled around it. The video heads "strike" the video information onto the tape in long, slanted scan lines creating a helical, or helix (screw thread), pattern on the tape (Fig. 1-2). The tape then passes over the audio record and synchronization record heads to provide all the information needed to store both picture and sound data. Finally, the tape is returned to the take-up side of the cassette.

With most helical-scan video recorders, the video signal is electromagnetically transferred to the video heads. Each head receives precisely one field of video information. One head records the information on the tape as one slanting track. The next video head receives and deposits the next field, and so on. Therefore, one slant-track, or one helical scan, equals one field, and two fields equal one frame or complete picture. Sync and audio are recorded by passing the tape over stationary heads. See Fig. 1-3 through 1-5.

A crucial part of the recording system is the *capstan* which, by rotating at a very precise speed (governed by sync signals on the

Fig. 1-2. The tiny video heads mounted on the head bar protrude from the head drum. As they rotate, they strike the tape and place a slanted track on the tape.

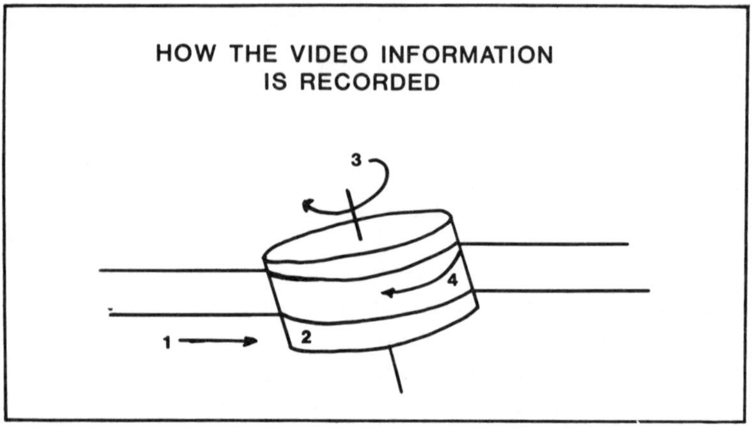

Fig. 1-3. The tape moves from left to right (1) and is wrapped around a cylindrical head drum (2) that is mounted on a slant and is rotating (3). Unlike earlier VTRs, the entire head drum rotates, and not just the heads contained inside the head drum. The heads mounted on the drum contact the tape and create a long slanted track on the tape (4).

tape), pulls the tape through the VCR. Variations in capstan speed cause instability in picture and sound quality.

PLAYBACK

To replay a tape the process is reversed. The video record heads become video playback heads and "read out" the recorded signal in the precise order in which it was put on tape. The TV monitor displays this playback video signal in exactly the same way

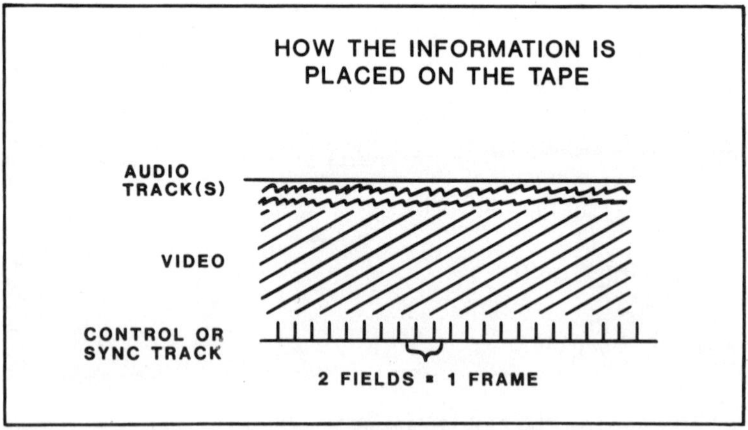

Fig. 1-4. The audio, video, and sync information are placed on the tape in this manner.

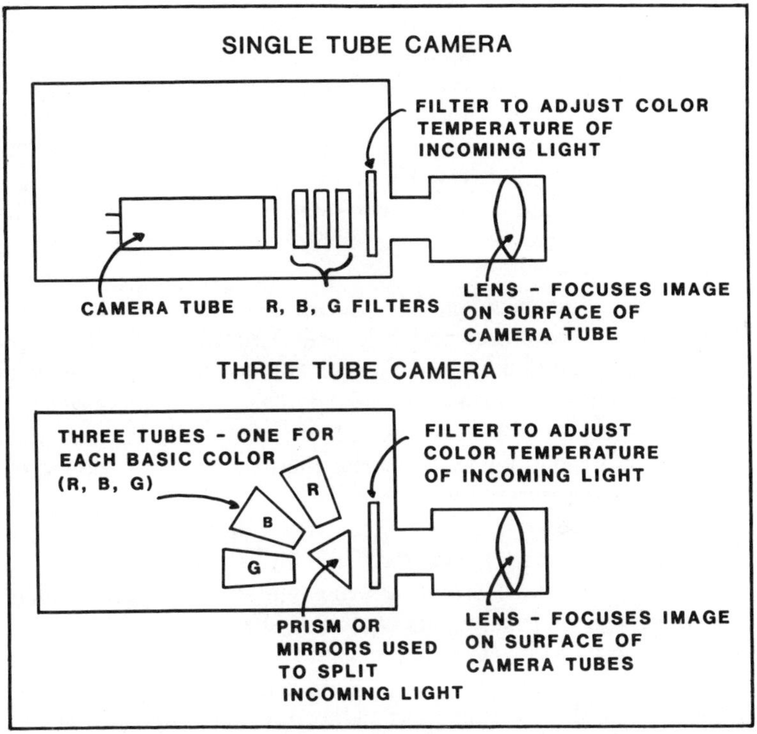

Fig. 1-5. This diagram shows the basic structure of a single tube and a three-tube camera.

it displays the camera signal. The rotating video heads "read" the prerecorded information stored on the tape and transfer that information into a fluctuating electronic gun in the rear of the television monitor's cathode-ray tube. The gun scans the phosphoric coating on the inside face of the TV screen. Fluctuations on the video signal elicit light of varying intensity from the phosphorus. It is these varying dots of light that, when seen all together, make up the light mosaic we call the television picture.

COLOR AND TELEVISION

A basic understanding of the television process is crucial to the effective application of color video production. But before I describe the color TV process, you first should understand the concept of color.

Those of us who are not in electronics or physics might use a dictionary's definition of color: something visible or recognizable only when rays of light fall upon it and distinct from its shape, size, or relative solidity or transparency...." In other words, to most of us color simply becomes another means of describing things or objects. This would be fine if we "saw" only in verbal terms. Early artists were faced with the problem of visually defining color. As the science of painting advanced, artists discovered—through the mixing and separating of different colors—that any color could be created by mixing one or more of the three primary colors: red, yellow, and blue.

With the advancement of color theory, many questions were raised. Does the object obtain its color from light or does light merely act as an illuminator? Physicists found that light travels in varying frequencies, or waves, and that these can be separated into the three primary colors mentioned previously. Colors could then be physically identified by measuring their wavelengths. So there's more to color than meets the eye!

The complexity that faced television engineers in trying to develop a color camera must have initially seemed overwhelming. They had to create a unit that could not only transform a light image into electrical information, but could also electronically identify and separate the varying wavelengths into television's physiological primaries of red, blue, and green. Television's primary colors vary slightly from those of paint (which are red, blue, and yellow).

DERIVING THE COLOR

The color system used in the United States is referred to as the NTSC system. The initials stand for the National Television Systems Committee; this committee worked with the Federal Communications Commission in formulating standards for the system. Engineers have evolved two basic methods for deriving NTSC color. The first, and more expensive, method requires three precisely matched tubes with filters placed in front of each tube. The incoming light image is focused by the lens and refracted with Dichroic mirrors, or prisms, onto the surface of the three tubes. The separation of red, blue, and green is accomplished by placing separate filters in front of each of the three tubes. One tube is used for each color.

The second, less expensive, and newer process for producing NTSC color uses a single tube that has, built into it, masks or filters that strip out or filter out the three basic colors of red, blue, and green. A fourth mask in some of the more advanced versions of this tube generates a "Y" channel (a luminous channel that aids in the remixing or blending of red, blue, and green information). It is very important in this type of system that the filters or masks be properly aligned at all times. Otherwise, inaccurate Y-channel information can be created, causing the color re-creation of flesh tones to appear too red or too violet. Depending on your system, the camera produces a picture signal consisting of red, blue, and green *or* red, blue, green, and Y-channel color information in addition to the electronically scanned video signal.

The third type of information that travels with the color signal is the synchronization or sync. The sync in a color signal performs the same function as it does in a black-and-white signal. Because we are dealing with additional information in a color signal, an additional electronic signal has been added to the sync signal. Its purpose is to assist in assuring the proper assemblage of this additional color information when it is reconstructed during playback. This additional sync signal is called the "color burst." The complete color signal (consisting of the video information, the color information, the sync, and the color burst) is then fed into the VCR.

COLOR PLAYBACK

Color playback is accomplished in essentially the same manner as in black-and-white playback. The difference is that a color TV has three guns to do the re-scanning—one for each color. In addition, there are three phosphors, matching R, B, and G, that make up the screen. Having these guns places a greater demand on the electronic system of the TV in assuring a coordinated re-scan process. One problem that evolves from the three gun re-scan is the overlapping of signals or *convergence*. Television sets periodically require adjustments for this overlap.

Color is a very complex subject. This chapter provides you with a simplified explanation of the color TV process. Color television technology is still a relatively new process, and new techniques are constantly being developed to make it better and less expensive.

Chapter 2

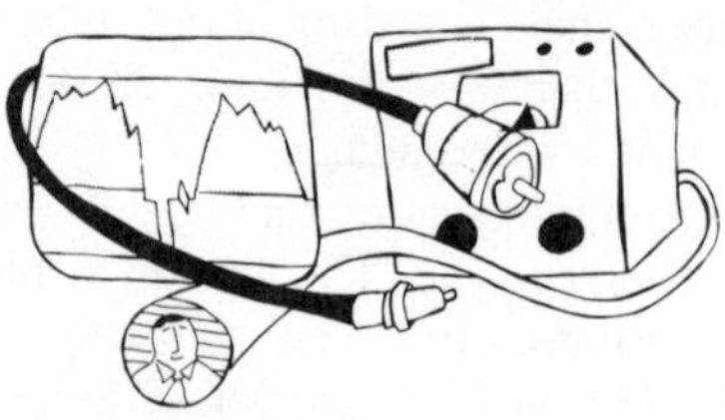

Controlling and Monitoring the TV Signal

Now that you have a basic idea of what makes up the TV signal, let's turn our attention to exactly what you can do with that signal. In producing videotapes, you generally will be working with three types of signals: video, audio, and rf. The rf, or radio frequency, signal is used in television to transmit audio and video signals through the air or through a single cable.

IMPEDANCE

When dealing with separate signals, it is important to recognize that audio, video, and rf signals are measured in terms of impedance. Different impedance levels are expressed in units of ohms. Very simply, *impedance* is the resistance a signal meets when traveling through a cable. For instance, a high impedance signal meets high resistance and, therefore, cannot travel very far without being reamplified. These signals can be interferred with by signals from CB radios, microwave ovens, fluorescent lights, or anything else that radiates signals into the air. The first step in controlling signals is to make sure that the proper impedance level is maintained and that the signal is protected or shielded from interference. Make sure the correct cables and connectors are selected to ensure proper shielding and impedance levels.

Pay attention to the impedance levels of the various inputs and outputs on equipment when hooking it up. Problems in this area will result in what is called an *impedance mismatch* that produces buzzing sounds in the audio and distortions in the picture—two real villains to the video producer. Once you have a system hooked up, one that maintains the proper impedance levels and provides protection from unwanted signals, you can do a number of things with it.

SPLITTING A SIGNAL

Feeding an rf signal out to more than one TV or an audio/video signal out of more than one VCR, requires the splitting of that signal. Inexpensive rf splitters can be purchased to split either a 300 ohm or 75 ohm rf signal. An audio signal can be split with something as simple as a Y adapter.

A video signal, on the other hand, requires some special attention. A video signal contains a very low amount of voltage (1 volt). As a result, this voltage level must be maintained. Therefore, when splitting a video signal, you must be sure that the voltage is properly reamplified to maintain its 1-volt level. See Fig. 2-1.

The splitters I have discussed thus far are referred to as *passive splitters*; they do not require external power. If you are setting up a system that requires splitting a signal more than once, an *active splitter* most likely would be used. Active splitters, or rf, audio, and video distributors can be purchased to provide the user with multi-

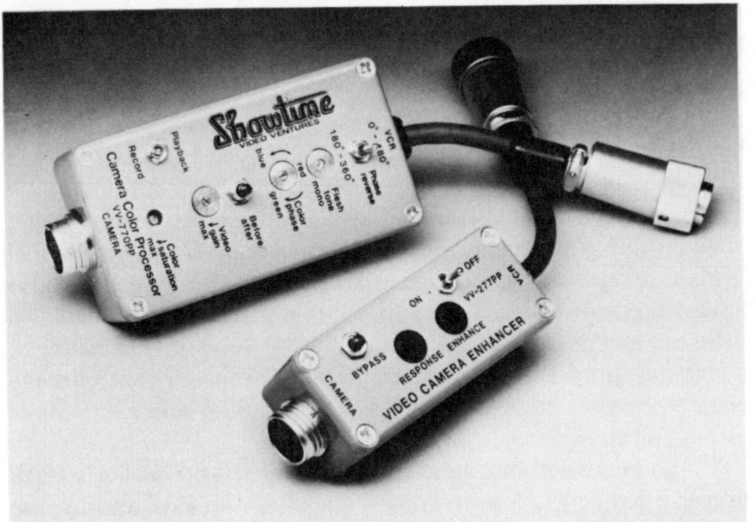

Fig. 2-1. Two inexpensive but very effective units that can be placed in line between your camera and your VCR to increase the overall picture quality (courtesy of Showtime Video Ventures).

ple outputs (3, 4, 5, 6, 7, 8, etc.). These units do require an external ac power source and they are usually isolated in terms of functions (i.e., rf splitter only, video splitter only, audio splitter only). More information on the controlling and processing of audio signals is provided in Chapter 3.

STABILIZING SIGNALS

Stabilizing a signal generally refers to dealing with the sync portion of the signal. You might have seen videotapes that have been copied using Copyguard. This is a duplication process used by many production houses to guard against unwanted duplication of their products. What copyguard does, in effect, is place a portion of the sync signal on the tape at irregular intervals. If you attempt to copy the tape you will find that your VCR will have a hard time locking onto the signal. "Copyguard" does prevent pirating of tapes but it also frequently creates problems during normal playback. TVs manufactured after 1979 can handle this problem and have little trouble locking onto the signal. Nevertheless, most VCRs, large screen TVs, and older TVs have difficulty handling the irregular sync. The result is a picture that rolls, jitters, jumps and breaks up. Video and rf stabilizers are sold that can overcome this problem. Their purpose is to provide a good picture for viewing but *not* for copying.

THE PROCESSING AMPLIFIER

You can affect the luminousness and color of the signal in a number of ways. Rf and video processing amplifiers or *proc amps* can be used to control and vary burst, chroma (color-saturation), and luminance of the signal. See Fig. 2-1.

With a proc amp, you can regulate the video level without affecting the sync signal. Remember that for VCRs with manual "video" level controls, increasing or decreasing the video level also affects the sync level. Often you will find that your sync level is a little too low or too high. When duplicating or editing tapes, try to maintain a proper video and sync level. If you can control these separately, you will assure better reproduction on later generations of the signal.

Another useful application of the proc amp is producing fades in prerecorded tapes. A prerecorded signal can be fed into the proc amp and, at the appropriate time, the video level can be turned down, producing a fade. The proc amp allows you to control the color burst and chroma levels separately, and this gives you a certain degree of control over the color make-up of the picture.

Signal enhancers are one way to create a "sharper" image. What happens is that the edges of images are amplified (either made darker or lighter). A good image enhancer can improve the overall sharpness of your picture. Within the last two to three years, a number of good, affordable processing amplifiers and image enhancement units have been produced for the home video enthusiast.

TIME BASE CORRECTOR

If you are producing tapes for CATV (cable television) or broadcast television, you will undoubtedly encounter the time base corrector (TBC). The signals produced on most ½-inch and ¾-inch VCRs generally are not very stable in terms of sync timing. They do not have to be because transmission and reproduction of these signals are usually very tightly controlled.

When a signal is to be transmitted through the air or over a cable TV system, the timing or rhythm of the sync becomes a crucial factor in assuring picture stability. Therefore the TBC is used to make sure the outgoing sync is timed correctly so—when the signal finally reaches the viewer's TV set—it is reproduced properly. Sync timing problems will produce pictures that jitter, roll, tear up, and hook or skew at the top. If you are going to produce programs for cable or broadcast, make sure the signal produced by your VCR can be properly fed into and accepted by the station's time base corrector.

Processing amplifiers, image enhancers, and time base correctors are useful pieces of gear. Keep in mind, however, that they are most successful when there is a good original signal. Remember, too, that a video or rf signal should not be processed or enhanced more than once. Consequently, it is important to determine the number of generations required for the production of a program before you ever turn on your camera.

MONITORING THE TV SIGNAL USING THE TV SET

There are a number of ways you can monitor your TV signal. This section addresses the video and picture portion of the rf signal. The monitoring of the audio signal is covered in Chapter 3.

The easiest way to monitor a picture signal when taping is to refer to your color TV set. Notice I say "color TV set" and not the camera viewfinder monitor. This is difficult when you are running around with your portapak. In such cases, it is wise to have a color TV set up (battery or ac powered) near where you are taping so you can periodically check the playback for color reproduction as well as audio quality. Because the viewfinder in your camera will be in most cases, black-and-white, it can confirm that you have a clear picture, but it cannot guarantee that the color is correct. When you are a production crew of one—and on the run—it is amazingly easy to forget to do a white balance, select the proper filter, or even choose the right f-stop—all of which affect the overall picture quality. If your camera has automatic white balance, a set filter, and an automatic iris, these can—for the most part—eliminate this type of problem. Even with all the automatic features, you really do not know what you have until you have seen it in "living" color.

When you playback a signal from your VCR on the TV, or when you are getting ready to tape off-the-air or from a broadcast rf signal, you should, of course, try to get the best possible picture. First adjust the fine tuning. Then adjust the color, brightness, contrast, and sharpness controls using a skin tone as your reference. If your camera has color bars, or if you have a color bar generator, use the color bar signal as a reference. I use my computer to create color bars. You will still want to check your skin tones because the color of a person's skin is what most of us refer to when watching TV.

Once you have adjusted the controls on your TV, leave them alone. If you find the color in the picture is changing when you check your tapes during a taping session, it is generally due to a change in the taping environment. Examples are a shift in the color temperature or the introduction of a different light source. The correction should be made at the camera (not the TV set).

USING A RECEIVER/MONITOR

Until now I have described using a TV set to monitor the signal. Another type of TV that is often used is called a *monitor*. A monitor is a TV set that accepts only audio and video signals. It has a panel with jacks on the back; the jacks let you feed a separate audio and video signal directly into the monitor. A monitor will give you a better picture because it lets you see the video signal (which is easier to isolate). An rf signal is more prone to interference. If you have a good television, however, and set it up properly, it should give you an adequate signal. Most monitors have an rf tuner built into them that lets them be used as either a monitor or a receiver. So you will also hear the term *receiver/monitor*.

Feeding separate audio and video signals into a TV projector unit will likely render a better picture than going rf. Keep this in mind if you plan to play a tape back on a large-screen projector. When hooking up the monitor, make sure the unit is properly terminated. The panel where the video signal is plugged in will also have a video out jack. Next to it you will find a switch. If the video out jack is not being used, it should be placed in the "75 Ω" or terminated position. Having the switch in this position prevents leakage of your signal. If you are using more than one large screen or monitor, this "video out" jack can be used to "loop" your signal to another monitor. Just make sure that the last monitor in the looping setup is terminated.

OTHER MEANS OF MONITORING THE RF AND VIDEO SIGNAL

There are several other ways to monitor your rf and video signal. Thus far I have focused on the monitoring of the color quality. It is also important to monitor the sync, video, and chroma levels. This is particularly true when you are going to be editing and duplicating your videotape or when your tape will be played back on cable TV.

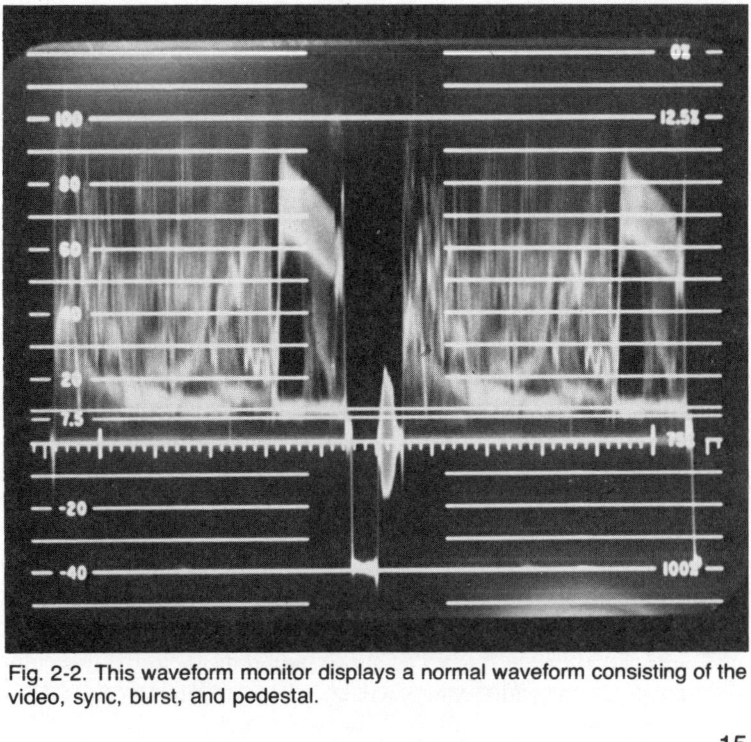

Fig. 2-2. This waveform monitor displays a normal waveform consisting of the video, sync, burst, and pedestal.

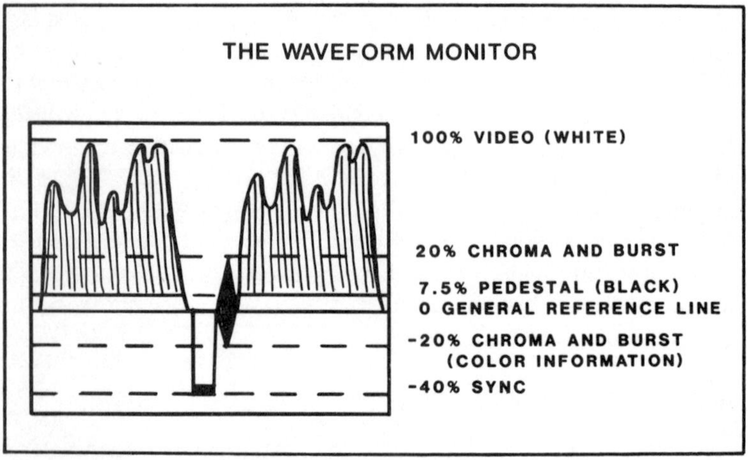

Fig. 2-3. The various information shown on a waveform monitor and the appropriate levels for each part of the waveform.

Automatic Gain Control

Most portable and nonediting VCRs have a built-in video *automatic gain control* (AGC). This should properly adjust the video, sync, and chroma levels, provided the camera is set up properly. Editing units, as well as many of the processing, enhancing, and distribution units now on the market, enable you to manually adjust the signal strength by turning a knob and watching a meter.

The Waveform Monitor

There is one drawback to using a meter: in most cases it gives you a read-out of the signal strength of the composite video signal. You sometimes need to see and separately adjust the video, sync, and chroma levels. The ideal piece of gear for monitoring these separate signals is a *waveform monitor.* Although you might never own a waveform monitor (the average cost is $2,000), you might sometimes have access to one if you are producing tapes for cable television or working with a client at a production facility.

The waveform monitor (see Fig. 2-3) displays the variations in signal voltage and sync pulse. These variations will directly affect the brightness, color, and stability of the picture. Starting at the top of the waveform monitor, the following are the signals that can be monitored and the measurement values that should be maintained:

Video Level. The white or amount of brightness in a picture is measured by the video signal. It should never go above 100 percent. When it does, the white values will be "clipped," making portions of

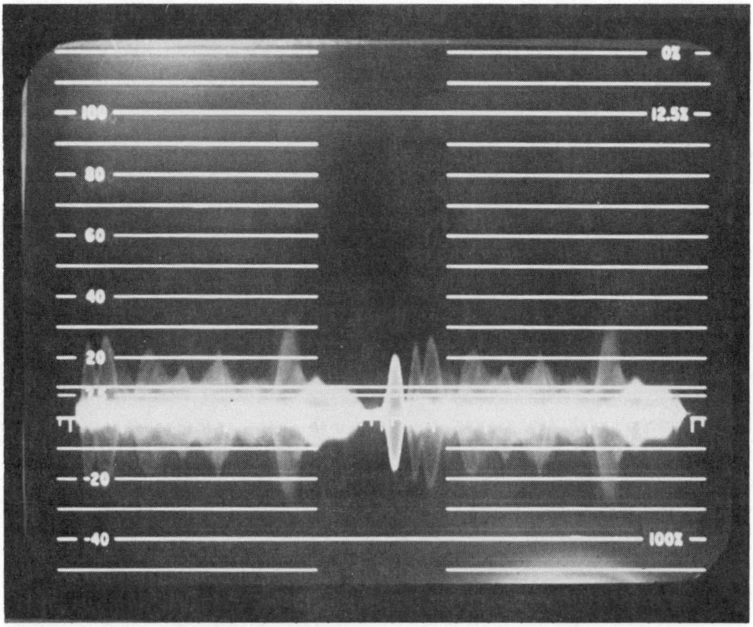

Fig. 2-4. By switching the waveform monitor out of its normal position, it will display the burst and chroma level.

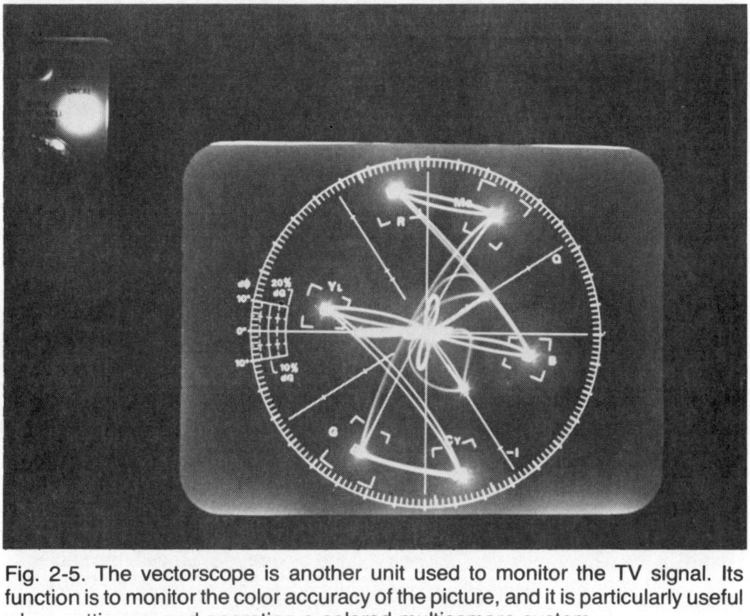

Fig. 2-5. The vectorscope is another unit used to monitor the TV signal. Its function is to monitor the color accuracy of the picture, and it is particularly useful when setting up and operating a colored multicamera system.

the picture too "hot." When a "hot" picture is recorded it will appear washed out. Contrarily, if the video level is too low, the result will be a dark picture that, when played back, will likely contain "noise" or have a grainy look to it.

Pedestal. The amount of blackness in the dark areas of the picture is measured by the pedestal level. It should be set at 7.5 above the general reference level. Increasing the pedestal will cause an overall loss in picture contrast. Decreasing the pedestal will create a loss in picture detail or sharpness.

Sync. The sync signal should be maintained at a steady −40. The sync signal is what provides the VCR with a reference for its automatic adjustments (i.e., video level). If it is incorrect, the recorded image will appear washed out or noisy. Also, problems can arise when you are trying to edit two signals with different sync levels.

Burst. The burst contains the information that provides the picture with color. It should be set to register with its top at +20 and its bottom at −20. If the burst registers greater or less, there will be resulting loss of color, color lock, or stability problems.

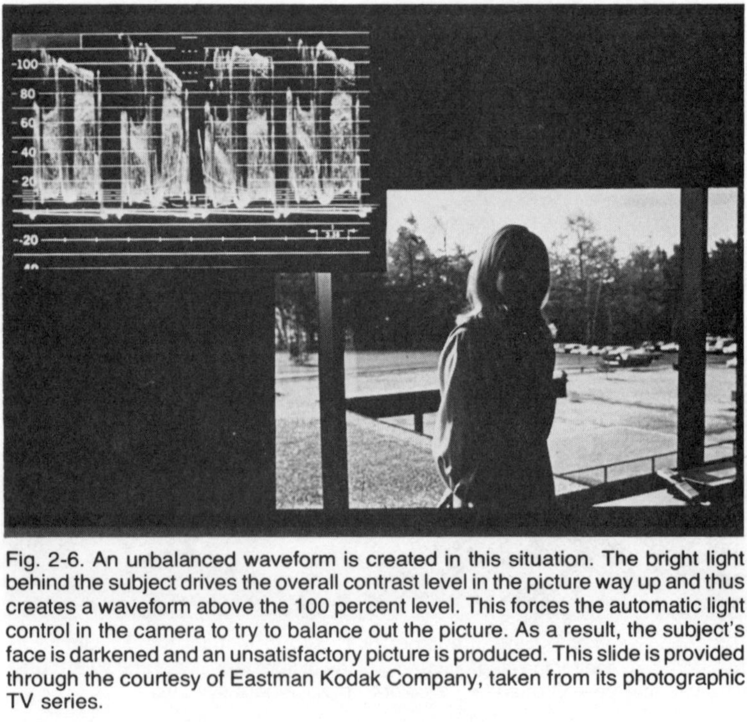

Fig. 2-6. An unbalanced waveform is created in this situation. The bright light behind the subject drives the overall contrast level in the picture way up and thus creates a waveform above the 100 percent level. This forces the automatic light control in the camera to try to balance out the picture. As a result, the subject's face is darkened and an unsatisfactory picture is produced. This slide is provided through the courtesy of Eastman Kodak Company, taken from its photographic TV series.

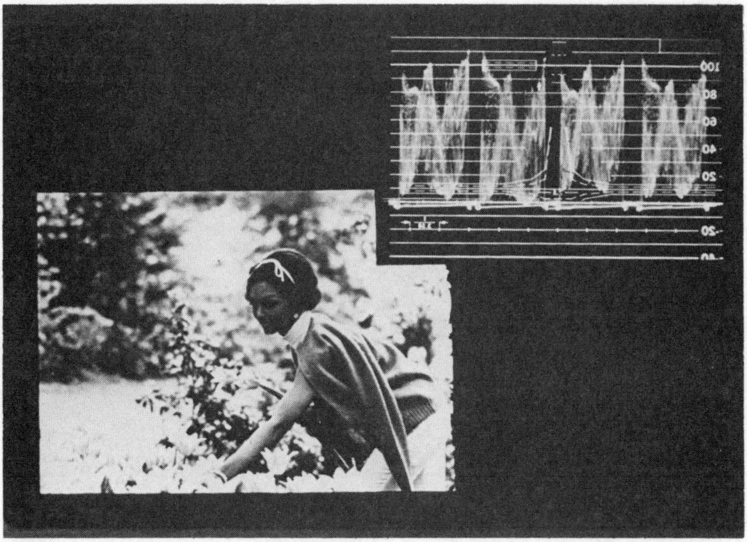

Fig. 2-7. An example of a picture that has an even contrast ratio that is evident in the waveform created. This slide is provided through the courtesy of Eastman Kodak Company taken from its photographic TV series.

Chroma. The chroma level is used to monitor color saturation. The top of the signal should never be more than +20 and the bottom of the signal should never be less than −20. Too high a chroma level will produce chroma noise that results in a grainy effect that only worsens with successive generations. Too little chroma level results in loss of color intensity.

The other signal you will often hear mentioned is the *horizontal blanking* or *vertical blanking*. Blanking signals are used in the rescanning process of a camera or receiver to blank out the retrace or rescan lines. Any tapes used on broadcast TV must conform to FCC regulations (which state that there should never be more than 21 lines of vertical blanking). Vertical blanking can be measured using the waveform monitor. See Figs. 2-2 through 2-5.

I have covered signal control and monitoring in fairly extensive detail in this chapter. I am not trying to make you into an engineer, but it is important to understand all these factors—particularly when we move on to the subjects of lighting, graphics, and editing. Spending a great deal of time planning and producing a videotape, only to have problems in duplicating it, can result in incredible frustration. A basic understanding of how a TV signal is controlled and monitored will prevent that kind of unhappy and exasperating experience. See Figs. 2-6 and 2-7.

Chapter 3

Sound Advice about Audio

It is impossible to determine which is more important in videotape production—the sound portion or the picture portion. Over 65 percent of all human communication has occurred via sound. We are surrounded by sound from the moment we are born. Our ears work even when we're sleeping. Consequently, sound is our most familiar means of communication.

My experiences indicate that if we watch a program on television and we lose the video (or picture), but retain the sound portion, it is likely our attention will still be held. If we watch that same program and lose the sound portion, we quickly lose interest. You can see from this example that, although we think of television as being primarily pictures, the video portion must work with the audio track to achieve effective communication.

It is easy to achieve crisp, clear sound when that's all you have to worry about. In a radio studio—where we cannot see what's happening—large, bulky microphones can be placed inches from a performer's mouth, and the room can be designed with acoustical panels and other backdrops that help record sound at its optimum. In television we are more limited. Because we must worry about what the audience *sees* as well as what it hears, we cannot use large, obstrusive microphones and acoustical backdrops. We need a sound source that is both unobtrusive and effective. Therefore it becomes more difficult to provide good sound in the television environment.

SELECTING THE MICROPHONE

The ideal microphone for small systems television applications would have these traits:

☐ It would be extremely small.

☐ It would be very rugged.

☐ It would be highly receptive to desirable sounds, yet reject unwanted noise.

☐ It could be used effectively for a single person or a large group.

☐ It would be wireless (for mobility), yet it would not be dependent on batteries or costly radio frequency transmitters.

Actually, no one microphone has all these features. Television has prompted the invention of several types of microphones that can be used in many different situations such as individual or group coverage, or outdoors in target-of-opportunity events. Therefore, we need to know something about the types of microphones available and how to choose the proper microphone for each particular taping situation.

The world of audio has created its own private mystique of terminology and applications. If you are familiar with what the basic terms mean, you will be much more effective in choosing the proper microphone for your system. There are five things to consider when selecting a microphone.

Frequency Response. Frequency response indicates the microphone's capability to reproduce sound over a wide spectrum (from the lowest bass tones to the highest treble squeaks). For the "pros," this frequency response is given a cyclic measurement called *hertz* (or Hz). You should know what ratings are applicable to quality professional-grade microphones. Choose a microphone with a frequency rating of approximately 80 to 13,000 hertz. This range will ensure optimum performance within the capabilities of your VCR unit.

Pick-up Pattern. The pick-up pattern is the microphone's sensitivity to sound that is generated in front of it, on the sides of it, or behind it. Pick-up patterns are generally one of two types: omnidirectional (the microphone picks up sound all around it) or unidirectional (the microphone is sensitive to sound in a very narrow width in front of it). You will also hear of the term cardioid pattern; it is a modification of the omnidirectional or unidirectional pattern. See Fig. 3-1.

Omnidirectional. If you were to inflate a large balloon and

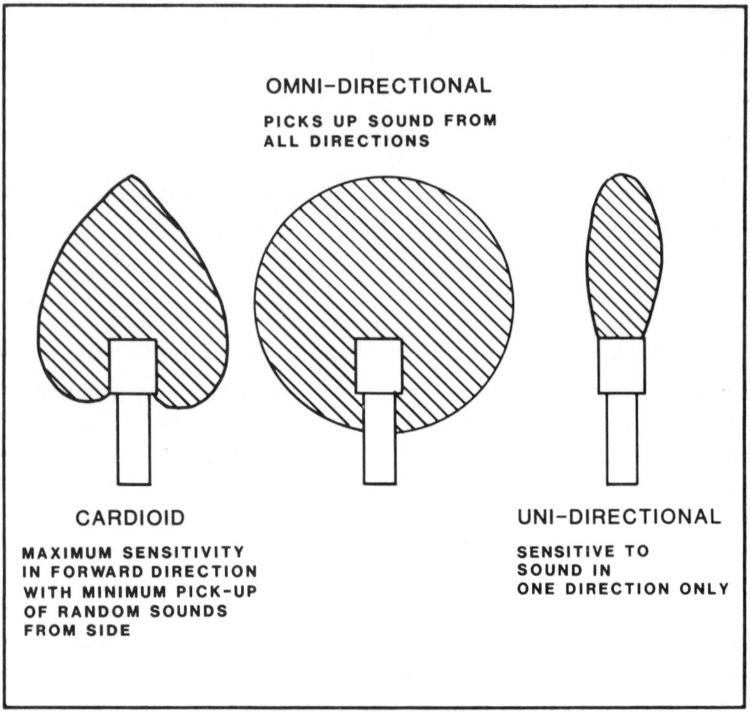

Fig. 3-1. Three basic microphone pick-up patterns.

suspend the head of our microphone at the very center of the balloon, you would notice a three-dimensional ring all around the microphone. This represents the pick-up pattern of an omnidirectional microphone. You could address the microphone from the front, the sides, or the rear and still have the sound effectively transmitted through the unit. Omnidirectional microphones are extremely effective in picking up sound from several people (such as a group seated around a table). They are more prone to feedback when used with a public address system and they tend to reproduce background noise such as paper rustling.

Unidirectional. Unidirectional microphones are highly directional in their pick-up pattern. They accept sound from directly in front of the microphone and reject sound from the sides or the rear. A good example of a highly specialized, unidirectional microphone is the "shotgun" type of mike used at sporting events. A person standing along the sidelines at a football game can point this type of microphone at the huddle 40 to 50 feet away and hear the plays being called by the quarterback—while the roar of the crowd is suppressed. Naturally, the applications of this type of microphone are limited; it might be hard to justify the price.

Cardioid. The cardioid pattern combines advantages of the unidirectional microphone (especially receptive to sound in front of the unit) and the omnidirectional (two sound lobes pick up sound on the sides). The cardioid pattern is being used more and more in television today because it works well with public address systems, it can be used for one or more persons, and it rejects unwanted background noise.

Ruggedness. Durability is the third characteristic to consider when selecting a microphone. If a lot of your video applications are portable, you will want a microphone that will withstand abuse. Shock mounting and wind screens (to prevent pops and gusts of wind from making noise, Fig. 3-2) are important for outdoor use. On the other hand, this type of microphone costs more. If you are working in a studio, less expensive (and less rugged) microphones are perfectly all right.

Presence. Obviously, a key to good audio is the microphone's ability to reproduce sound accurately. Part of the task of good sound presence or reproduction is governed by the frequency response. The other part is determined by the method used to capture and reproduce sound waves. Good presence is achieved in different ways with different types of microphones.

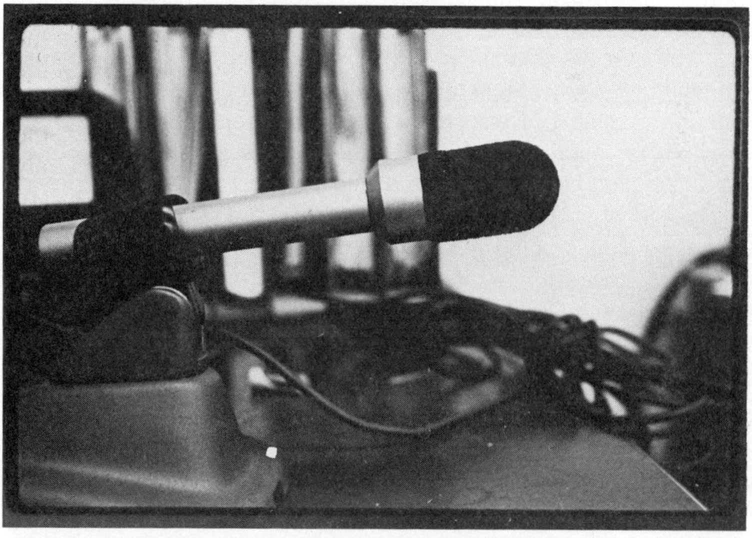

Fig. 3-2. This desk-stand mounted microphone has a wind screen on it to cut down on "pops" caused by talking into it.

For example, the *dynamic microphone* is basically a small loudspeaker designed to work in reverse. It consists of a magnet, a coil, and a diaphragm. The coil moves and produces an inductance. In turn this produces a signal that can be directly fed into an amplifier. The favored frequency in a dynamic microphone usually lies in the middle of the audio range.

The *crystal microphone* is like the dynamic microphone only it uses a small crystal rather than a diaphragm to react to the incoming sound waves. Its frequency range is much narrower than that of the dynamic microphone (which makes it unsuitable for use in video production work).

The *ribbon microphone* uses a metallic "ribbon" that has been specially stretched and calibrated (enabling it to accurately reproduce sound vibrations). Its big disadvantage to video users is its size and sensitivity. It does not like to be moved around.

The *condenser microphone* has two main elements. A condenser element reacts to sound waves and transmits them to a coil and an oscillator circuit amplifies the sound waves. This type of microphone can reproduce and transmit a wide frequency range of sounds. Its current is normally provided by a power pack built into the microphone cable that is 5 to 6 feet from the microphone.

The *electret condenser microphone* (Fig. 3-3) is just like the condenser microphone, except that it is smaller and more compact. Its power is supplied by a battery that is usually contained in the microphone.

The *wireless microphone* (Fig. 3-4), though very expensive at the moment, should soon be affordable for the home video user. It contains a small FM transmitter that is either built into the microphone or contained in a small box that can be placed in the pocket or attached to a belt. The transmitter has a short antenna attached to it. The FM signal is picked up with a special FM receiver that is set to the frequency being generated by the transmitter. The signal from the transmitter is then fed to the VCR or mixer as a line or microphone signal. Though highly satisfactory for the most part, wireless microphones do have these disadvantages:

☐ They are more susceptable to interference by other signals in the air such as walkie-talkie and CB transmissions.

☐ As long as you have a clear line of sight, the FM signal can usually be picked up clearly. If you transmit through walls, there might be problems.

Impedance. Impedance is the magic, mystery word of the wonderful world of audio (perhaps justifiably so). If you talk to 10

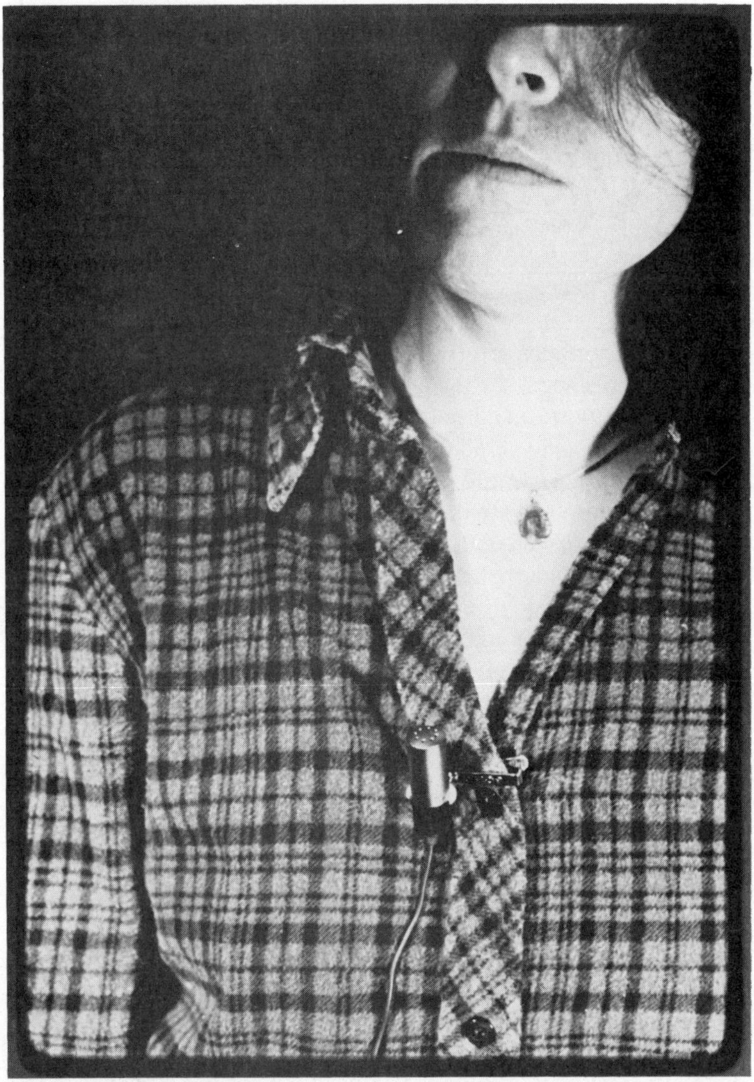

Fig. 3-3. The electret condenser lavalier microphone is a very popular microphone. Be careful when clipping it to your subject. Don't let it rub against clothing or get covered up by mistake.

engineers about impedance, we will get 10 different definitions of the term. The word is one that leads up many alleys, depending on the specific audio term to which it is applied. For our purposes, we can think of impedance as meaning the resistance of the microphone to the transmission of sound.

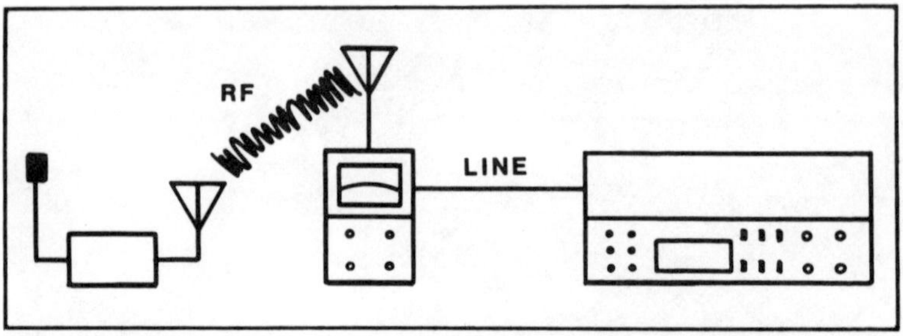

Fig. 3-4. This diagram outlines how a wireless microphone is set up.

Basically, there are two kinds of impedance: high impedance and low impedance. In the early days of radio, high-impedance microphones were used because they matched the natural high-impedance property of radio grids. As time went by, it became easy to design more versatile electronic equipment. High-impedance microphones are easy to design with the good properties, but they are limited in the amount of cable run we can use. If you use a high-impedance microphone with a cable over 12 to 15 feet long, you will find it quickly begins to lose signal response. In comparison, low-impedance microphones give the advantage of much longer cable runs with little or no loss in signal response. Because many microphone applications require long cable runs, the low-impedance type is much more advantageous. The majority of VCR and audio manufacturers in today's market have designed their equipment to accept low-impedance microphone inputs.

Microphone impedance is rated in ohms. Most low-impedance microphones have a rating of 200 to 300 ohms. The high-impedance microphones will be rated at 10,000 ohms or more. You can see there is a broad dividing line between microphones of high- and low-impedance formats. For general information, anything that is 3000 ohms or less is low impedance and anything above 3000 ohms is high impedance. Some video tape equipment, such as the portable units, are designed to give an average impedance input. This makes the equipment compatible with either high- or low-impedance microphones. If you have a low-impedance microphone and a high-impedance VCR, or vice versa, you can buy a *line transformer* that can change the impedance of your microphone. If you have a low-impedance microphone, connect the line transformer to the end of it and it converts the signal to high impedance for your VCR. Most of these transformers can be used both ways so your high-impedance system can be converted to a low-impedance signal by switching the transformer around.

One final note on selecting microphones is that you don't have to go broke buying adequate microphones. A good professional-quality microphone with the characteristics I've discussed can be bought for between $30 and $80. It's a good idea to buy a variety of microphones. Obtain some to use in hanging applications, some to be used around the talents' necks or attached to their clothing, and perhaps some for special uses, such as the FM-wireless type. See Fig. 3-5.

HIGH-LEVEL SIGNALS

Microphones generate either a high- or low-impedance signal. The microphone plugs into your VCR at the mike input. In examining the other jacks on your VCR, you will find an audio or line-in jack as well as an audio or line-out jack. These line input and output jacks are used for high-level signals. High-level line signals are electronically different from high- and low-impedance signals. High-level signals usually come from stereos, other VCRs, public address systems, and audio mixers.

The high-level line input jack should be used when feeding audio into your VCR from an audio tapedeck, public address system, or from another VCR when dubbing a tape. If you want to copy the sound portion of your tape onto an audio tape or cassette recorder, feed the audio to the tape or cassette deck using your high-level line output. Naturally, in order to do this, your audio gear needs to have a matched input or high-level line input capability.

MICROPHONE MIXERS

Microphone mixers are available in all sizes and shapes from simple, battery-operated units to complex, professional theater systems. Some mixers act as *combiners*; they take the sound information from two or three microphones and combining it into one output for the VCR. Other units are designed to reamplify signal strength to keep it as close to the original level as possible. You can mix microphone inputs and high level, amplified auxiliary or line inputs into one common output. You can also find mixers that will take up to 20 inputs, add reverberation, echo, dolby B noise reduction, compressors, and other features into the system. Regardless of how simple or complicated the mixer, the technique for using it is always the same. See Figs. 3-6 and 3-7.

Let's assume you hooked up your microphones, checked them out, and connected them to a mixer that is connected to a VCR. Next, move all controls to the closed or zero position. Take the

	CRYSTAL	RIBBON	DYNAMIC	CONDENSER	ELECTRET-CONDENSER
FREQUENCY RESPONSE*	poor	excellent	good to excellent	good to excellent	good to excellent
SENSITIVE TO HEAT & HUMIDITY	yes	yes	some	yes	yes
ABILITY TO WITHSTAND ABUSE	no	no	yes	no	no
PORTABLE	yes	no	yes	yes	yes
POWER REQUIREMENT	none	none	none	yes battery	yes battery
COST	$5.00	$300.00 +	$50.00 +	$50.00 +	$50.00 +

* when mic is used to reproduce the human voice

Fig. 3-5. Microphone design categories.

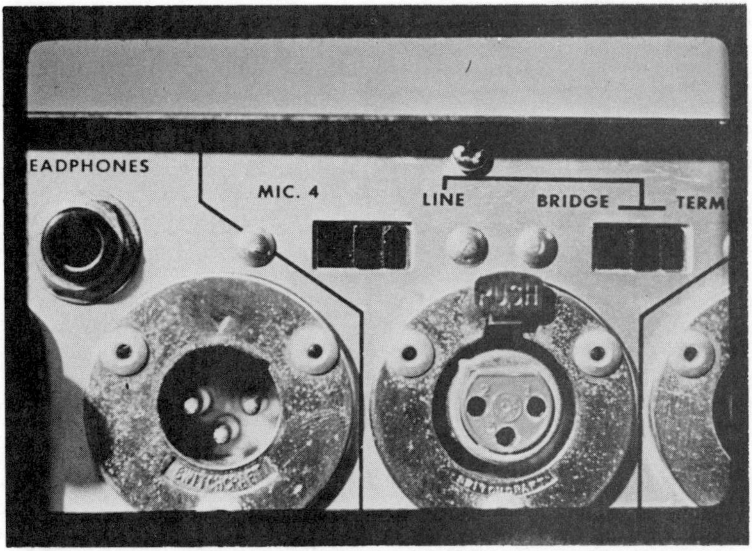

Fig. 3-6. One of the inputs on the back of this microphone mixer can be switched to accept either a line or microphone input.

master output control and rotate or slide it to the closed or zero position. Take the master output control and rotate or slide it to the seven or ¾ open position. Fold some blank 3-×-5-inch cards and tape them above the individual channels or pots on the mixer. (*Pot* is audio jargon for potentiometer.) Above each pot, mark the name of the person who is the primary sound source for that microphone. This will prevent you from accidentally closing the wrong microphone later on in your production. See Fig. 3-8.

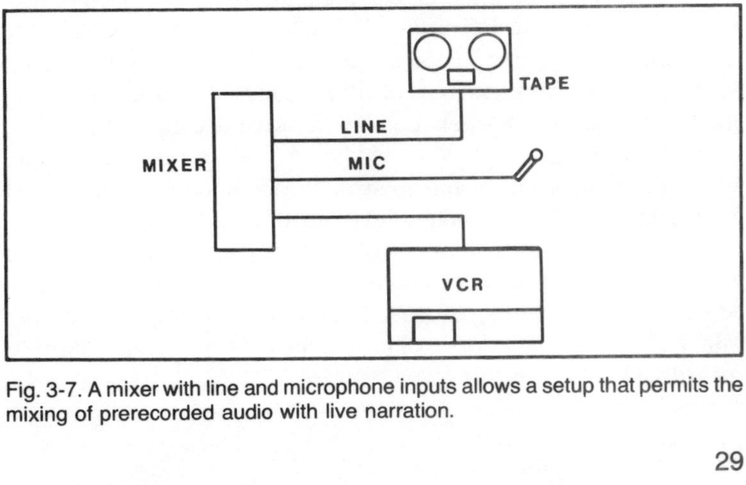

Fig. 3-7. A mixer with line and microphone inputs allows a setup that permits the mixing of prerecorded audio with live narration.

Fig. 3-8. Identifying the various pots on the mixer will prevent accidental fade outs.

Zero all pots and—with the master volume control open to ¾ position—put your hand on the #1 pot for your first guest or talent. Ask the person for his name, address, place of business, etc. While he is speaking, rotate and slide the pot open; watch the VU meter on the mixer of your VCR until a good healthy swing into the "normal" range is indicated. Occasional peaks into the red band are OK, but try to keep about 80 percent of your information in the black ("normal") area. Note the level you have set on the first person's microphone. Write it on the card above his channel or pot and then return his pot to zero. Repeat this procedure with second, third, and fourth persons until you have noted each level for each individual pot. It's important to close each pot after each sound source check.

If your VCR has a choice between manual or AGC audio, it is best to use the manual position. This gives you more control over the sound and, on some VCRs, keeps the VU meter operable during recording. If you use the AGC position, you might notice considerable background noise when there are lulls in the conversation. The volume control on the VCR should be adjusted to the sync and the VU meter with the one on the mixer. If the mixer does not have a VU meter, then adjust your sound level strictly on the VCR.

The reason the master control is initially set at the 7 or ¾ open position is to give us some control to turn the overall level up or down once the recording process begins. Most people will be very calm and give you a normal volume check before the tape starts to record.

Once the lights are on and the machines are running, their adrenalin level seems to increase, causing their volume to rise accordingly. If the overall tone of the production becomes louder, you can decrease it with the master volume control. On the other hand if a individual participant's voice becomes louder or softer, you can control his microphone with his specific pot—without affecting the others.

If you are premixing prerecorded music through a line to provide background for opening and closing titles, remember your narrator's voice. Keep the music volume proportionally lower on the VU meter when you do your checks and, of course, use your test recording to check the need for any volume corrections.

You should have a good headset to monitor the sound. Often there are numerous distractions in production situations; having a headset will help focus your attention on the sound. See Fig. 3-9.

There are many excellent, small microphone mixers at affordable prices on today's market. Those that have their own ac power cord are generally more dependable than those that operate on batteries. It's also a good idea to choose a mixer that can expand as your system expands. Many of the professional-quality mixers available have a mix option that will allow you to stack mixers if you need four, eight, or twelve inputs. You can connect different mixers through this mix feature and continue to expand your system by stacking mixers until you can no longer reach the box on the top row! Choose a microphone mixer and its features carefully; much depends on your applications, your studio or room acoustics, and the number of inputs you want.

CHECKLIST FOR RECORDING AUDIO

Here is a checklist you might want to copy and tape to the top of your microphone mixer.

☐ Microphones connected, individual pots zeroed, master volume open to 7 or ¾ position.

☐ Mark and identify sound source for each pot.

☐ Check the microphone on pot #1. Note the value. Return the pot to the zero position. Repeat the process for each sound input.

☐ Select the manual audio control position of the VCR and place the VCR in record. Adjust the VCR audio gain level to correspond with the indications on the mixer. If your VCR has no manual gain control, you'll have to use AGC. Manual is preferred.

Fig. 3-9. Headphones enable the camera operator to closely monitor the audio during taping sessions. Not only does this enable him to better follow the action, but it also gives him a means of detecting problems that could go unnoticed until playback when it might be too late to do anything.

☐ Open each pot to the prechecked level. Close the master input.

☐ With the VCR in the record mode, open the master control and cue talent.

☐ Always make a test recording before you begin serious taping.

AUDIO DUB

Adding sound to prerecorded video is a relatively easy production technique that usually provides excellent results. This is particularly true when you are trying to "salvage" boring or uninteresting video. It is amazing what a little music, some well-placed remarks, or some simple sound effects can do! The technique for adding new sound to prerecorded video is referred to as *audio dubbing*.

A number of factors will affect the audio dubbing technique you use; the key one is the number of audio channels in your VCR. I will first discuss audio dubbing on VCRs with one audio channel.

When the audio dub button is engaged with the VCR in the playback mode, it means that the original audio on your tape is being erased and that a new audio signal is being added. The new audio can be fed into your VCR through the microphone input or the high-level line input. As when adding original sound, try to always go direct. In other words, don't use a microphone in front of a stereo to pick up music. Use the line output on the stereo and feed it directly into your VCR's line input. By using an audio mixer, you can combine line and microphone inputs that create very professional results.

In situations where you want to add a narration with music, try to prerecord using a two-track audio tape recorder. Record the narration and then the music. This way you don't have to worry about riding your audio levels for the narration and music during the actual audio dub onto your videotape. One note of caution is that if your VCR has audio AGC, make sure you do a series of test recordings so you end up with the proper voice and sound mix. Sometimes when you feed a "mixed" signal into a VCR that has AGC, you will find your background audio (music or sound effects) should be recorded at a fairly low level. Otherwise, the VCR's AGC will "boost" it or attempt to equalize it to the level of the primary sound source.

If you are inserting new sound, you have to determine an "in point" and an "out point" for the audio dub. Look for a visual action cue for an audio "in" cue point or point at which to start the audio dub. If you introduce a new narration, time the length of the segment, using a stopwatch, down to your audio out point. Record the new narration on an audio tape recorder. This way you know the length of the audio insert will be correct. If you are adding music, you can usually rely on an audio or visual action cue as your in point and a visual action cue as your out point. I seldom use the counter because I have found it a very inaccurate source for determining either an in or an out point.

VCRs with two audio channels provide much more flexibility. They allow you to use a number of audio mixing and dubbing techniques. If you use channel one to record a person making a speech and decide to add music later, you can use one of two basic methods: the mix mode or the mix-up mode.

The Mix Mode. By placing the VCR in the channel mix mode and activating the channel 2 audio dub, you can dub your music on to channel 2. If you use this method, remember the following:

☐ Be sure the tape is played back in the mix mode.

☐ Prior to taping the speaker, be sure to leave a channel open for the audio dub.

☐ If you copy the tape and feed the audio from channels 1 and 2 out, using the monitor output, you might encounter a level mismatch when using a single-channel, automatic-gain-control VCR; if so, feed channels 1 and 2 out separately into a mixer and regulate the individual columns.

Mix-up. Let's take the same tape and this time use the audio mixer. First feed the audio from channel 1 and then the audio from our tape recorder or stereo into the mixer. The mixer's output will then be fed into channel 2 on our VCR. When we play back the tape and put channel 2 on our VCR into the audio dub mode, you will record both the speaker's voice and the music on channel 2. If your VCR has the capability to audio dub on either channel 1 or 2, you can mix back and forth several times, each time adding sound. This is a process frequently used in broadcast TV (particularly in news production). Keep in mind, however, that each time you go from one channel to another you will lose some quality. Making the effort to get good original sound will greatly affect the quality of your audio dub when using this method.

BLANKETING REVERBERATION

When going on location, always take some old blankets or packing pads to protect your gear in transit and to lessen room reverb. By draping blankets over a chair or hanging them from a wall, you'll be surprised at the reduction of reverberation. If you find yourself in a room with wall heating and cooling units that cannot be turned off during your taping session, packing pads can be thrown over the units to "blanket" out their disturbing clatter.

PROCESSING SOUND

It must be clear to you now that the audio signal, like the video signal, is only as good as it is! Countless times people come to professionals bearing tapes with hums or buzzes and saying, "Well, can't you just process it out?" The answer is no. Sometimes you can hide it, but the unwanted noise will always be there. And it just becomes worse as you copy or edit it.

"Processing" audio refers to about everything from mixing different sound sources to adding reverb or other special effects. Most of this is done with special audio processing gear. A visit to your local audio store will give you a good idea of the types of gear available (as well as the effects that can be accomplished).

Start simple. After equipping yourself with the necessary microphones, the next piece of gear you will want to add to your system will be an audio mixer. Make sure the mixer you select accepts both microphone signals and line-level signals.

The next piece of gear you add should be some type of tone control or equalizer unit. Using this unit either during production or in post-production editing or duplication, you can affect the overall tone of the audio signal. A good equalizer will allow you to affect separately the high or low frequencies you don't like, and remove or "filter" them out of your sound track without disturbing the needed frequencies. A good example of this is the removal of the rumble created by wind hitting a microphone. By increasing the treble control, you can filter out *most* of the low-frequency rumble created by the wind. Notice I said *most*. Chances are, if you had used a wind screen on the microphone in the first place, you wouldn't find yourself trying to correct this type of problem!

Your mixer or equalizer unit might have a low-cut filter switch. The low-cut filter will cut out low frequencies and is especially good to use when you are recording a person speaking. It will also help in filtering out room reverb or echo. The low-cut filter should not be left on when you are recording music. Music will produce both high and low frequencies you will want to record.

PLANNING YOUR AUDIO PRODUCTION

The recording of good audio requires a lot of skill and practice. In setting up your audio production, here are some key factors to consider:

- ☐ What are you going to be recording?
- ☐ How large a group will it be?

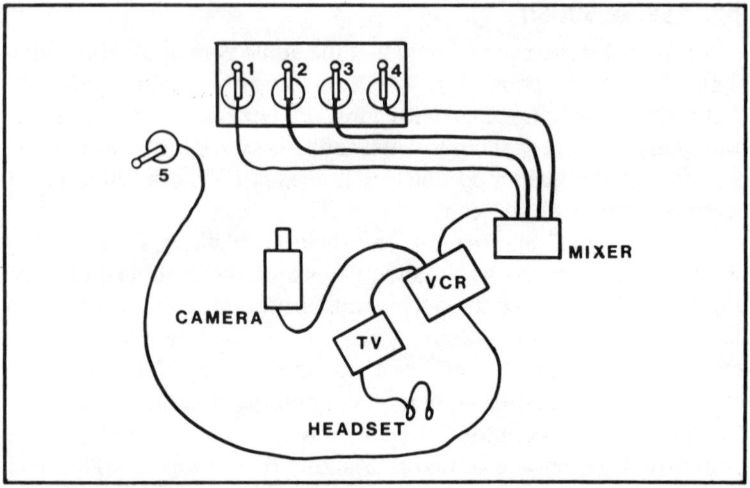

Fig. 3-10. Audio Production Situation: You must cover a four-person panel discussion; the panelists will be taking questions from the audience. You have five microphones, but your mixer will only accept four microphone inputs at one time. Solution: Four desk-stand mounted cardioid microphones (1, 2, 3, 4) are used for the panel. A stand mounted cardioid microphone is set up and designated as a question station to which persons in the audience wanting to ask questions must come. Microphones 1, 2, 3, and 4 are fed into the mixer. The mixer's line-out is fed into the VCR's line in. Microphone 5 is plugged into the VCR's microphone input only when it is used to ask a question. When a microphone is plugged into the VCR's microphone in jack, the line input is automatically cut off. The audio is monitored using headsets plugged into the TV. Before using this production technique, test it out on your system.

☐ How close can you get to your subjects?
☐ How much will your subjects be moving around?
☐ What is the recording environment like?
☐ Will you be adding narration, music, or other sound effects to the original audio?
☐ Is the tape going to be edited or duplicated?
☐ What kind of system will the tape be played back on?

PRODUCTION TIPS

The following are production tips I have picked up over the years. I have found them to be very helpful in planning and producing the audio portion of video productions. See Figs. 3-10 and 3-11.

Make sure all cables are properly secured. Use gaffers tape to cover and tape down cables that are strung across the floor.

Tie the microphone cable (or cables) to a table leg to prevent dumping the mixer or VCR on the floor in case someone trips on the cord. Also, tie or tape the microphone cable to the mike stand.

When using a lavalier microphone, clip, tie, or tape the cable to the user's waist if possible. This will help relieve the stress on the microphone in the event the user ventures too far for the cable.

When stringing long cable runs, tape all cable connections to prevent them from being accidentally pulled apart.

Before a taping session, warn subjects about handling the microphone cable. This can cause distracting vibrations that will be picked up by the microphone.

If you use a desk stand microphone on a hardtop surface, set it on a ¼-inch piece of foam rubber. This will help absorb unwanted vibrations created by people hitting the table or tapping their fingers on the table surface.

Buy a headset so you can monitor the audio during recording and playback. I use a headset with a long extension cord that enables me to move back and forth between the camera and VCR during the taping session. I always listen to the audio.

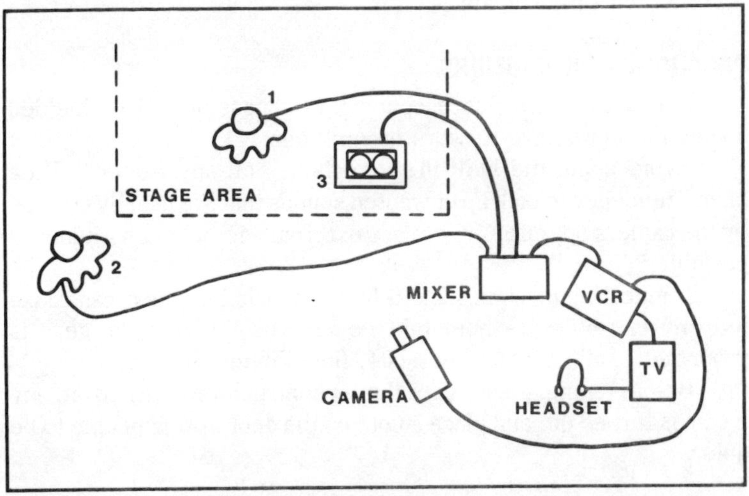

Fig. 3-11. Audio Production Situation: An individual is going to give a lecture. He'll be moving around on a stage and playing back short musical pieces using an audio tape recorder. He'll be taking questions from the audience. Unfortunately, the seating arrangement is not suitable for the setup or for a question station. Solution: A lavalier microphone (1) is placed on the lecturer, and a person with a shotgun microphone (2) is positioned in front and to the side of the lecture area. The line output from the audiotape recorder (3) is fed directly into the mixer's line input. The audio is monitored using headphones that are hooked up to the TV.

Always make a test recording. If a problem in the audio develops during the taping session, stop the session and correct the problem.

It is always good to get extra background sound (especially when taping on location). I usually either let the tape just roll at some point to record extra audio or I take along a cassette recorder and use it to record ambient sound.

Tell your talent to talk across the microphone (not into it).

The small batteries used in electric condenser microphones should be removed when not in use. Clean them periodically, use a pencil eraser to scrub the two surfaces.

Have a box of emergency connectors and adapters that you carry to every location assignment. Two things I carry are a box of minialligator clips and a small bottle of rapid-bonding adhesive (alpha cyanoacrylate) contact cement. This combination is just what's ordered to jiffy-rig a replacement for a lost tie-clip mount on a lavalier microphone. Rapid-bond cement is strong enough to bond your fingers. Keep a small bottle of solvent (nail polish remover or acetone) with the cement.

PRODUCTION PRECAUTIONS

Stay away from fluorescent lights if possible. Use shielded cables if you do have to work around them.

Avoid using the built-in microphones on any camera. They have a tendency to pick up unwanted sounds that are usually created by the camera operator. Try to use external microphones whenever possible.

Duplicating machines and telephone switching gear can cause intermittent 60-cycle hum line noise. Make sure your gear is grounded. If the problem persists, find a different ac outlet.

Before taping, make sure all telephone bells are turned off, any music is turned off, and place a note on the door asking people to be quiet.

Chapter 4

Lighting Methods and Techniques

Many video users initially avoid supplemental lighting gear because it gets in the way and, more often than not, causes new production problems. It creates shadows and it gets into the subjects' eyes. Nevertheless, the more you use your video system the higher your standards will become. You will find yourself paying more attention to improving picture clarity and achieving good color. A basic knowledge of the technical and aesthetic requirements for TV lighting will vastly improve the overall picture quality and color of your productions. A short period of time spent learning lighting fundamentals will be well worth it. See Figs. 4-2 and 4-3.

COLOR TEMPERATURE

Two technical requirements you should understand are *color temperature* and *light intensity*. All sources of light have a color temperature rating measured in Kelvin degrees (K). A candle, for example, produces about 1600 degrees K. An incandescent light bulb produces 2700 K. The light outside on a bright, sunny day is approximately 8000 degrees K.

Our eyes automatically compensate for different color temperatures. In other words, the color of an object appears the same both inside with incandescent light and outside under sunlight. This is not true with a video camera. If you videotape an object first inside and then outside, that object will not be the same color in both pictures (assuming you don't make any adjustments to the camera).

Fig. 4-1. Both sunlight and quartz light are being used to light this production setup. Because the production is being done with a black and white camera, the mixing of different light sources makes no difference. If a color camera were being used, the presence of two different color temperatures, that of the sunlight and quartz light, would produce a reddish or bluish tone and would have to be corrected by using colored gels. In this case, you would probably use a gel over the quartz light to correct its color temperature to that of sunlight. Your photographic store can supply you with a chart that will list the different gels used to correct quartz and other lights to that of sunlight and vice versa.

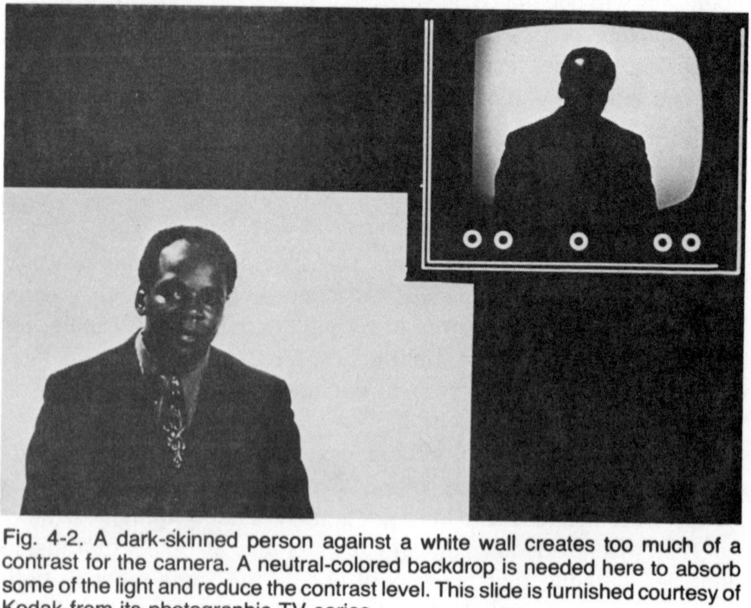

Fig. 4-2. A dark-skinned person against a white wall creates too much of a contrast for the camera. A neutral-colored backdrop is needed here to absorb some of the light and reduce the contrast level. This slide is furnished courtesy of Kodak from its photographic TV series.

A TV camera has this problem because it was designed to provide the correct color of objects illuminated with a light source having a color temperature of 3200 degrees Kelvin. Most cameras have a filtering system that enables the camera to compensate for different color temperatures.

The key factors here are to first make sure the correct filter adjustment has been made and, second, to white balance the camera before the taping begins. An easy way to spot problems is to note the overall color "tone" of the picture. A picture with a bluish tone means that the color temperature in the environment is higher than 3200 degrees. A picture with a reddish tone means it is under 3200 degrees Kelvin. In either situation, make sure you are using the correct filter and perform a white balance.

The best inside lights are quartz, iodine, or tungsten-halogen lamps with a color temperature of 3200 degrees K. These lamps have been specially designed to produce this color temperature consistently over a long period of time. Several precautions should be observed when using these lamps:

☐ Use a cloth or gloves when handling these lamps. Touching the lamp with your fingers can create grease spots on the crystal. These spots can cause an uneven distribution of heat and create a bubble that could swell and burst, destroying the lamp. If you do accidentally touch the lamp, wipe the fingerprints using a tissue and some denatured alcohol.

☐ When turning on the light, always point it away from yourself and bystanders. Damaged bulbs can explode (sending glass particles into the air).

☐ Let a light cool off before moving it.

LIGHT INTENSITY

The second technical requirement to consider is the intensity of the light in the taping environment. The intensity of light is measured in units called foot-candles. As you might have guessed, this measurement is based on the intensity of candle light measured a foot away. A variance in the intensity of light in a given environment produces contrast with that environment. The TV camera has a problem handling contrast. Consequently, a key factor to keep in mind is that an evenly lit environment will produce a clearer video picture. Likewise, controlling the intensity of the light in the environment allows you to control the contrast. There are several ways to do this.

Wattage. The higher the wattage the higher the intensity. Quartz-iodine lamps usually come in either 650-watt or 1000-watt configurations. In choosing what wattage to use, consider how close the lights will be to one another and to the subject. Also consider how much power you will have. Use this formula to make sure the available electrical supply can support the required lighting units:

$$\frac{\text{POWER (WATTS)}}{\text{VOLTAGE}} = \text{AMPS}$$

Thus, a 1000-watt lamp could be used on the circuit because most household circuits carry 15 amps each.

Barndoors, Scrims, and Gels. Several items that can be attached to a light and can be used to control its direction and intensity. Be careful when using color gels; they can cause a change in the color temperature environment. See Fig. 4-3.

Dimmers. Most TV studio systems are set up with dimmers or rheostats that can be used to raise or lower the wattage on lights. This is a convenient but expensive way to produce even lighting.

Camera Controls. The opening and closing of the iris on the camera can increase or decrease the overall light intensity. Likewise, diffusion filters placed on the front of the camera lens can soften the intensity.

MONITORING LIGHT INTENSITY

There are a number of ways to control light intensity, but how do we know when the intensity level is correct? The first method is to look at the picture on your TV before setting up lights. Try to spot areas of high contrast such as a window in the background, the reflection of a polished floor, or a white wall. Then look for areas of low contrast such as dark draperies or tablecloths or dark clothing. You can correct these problem areas by changing the camera angle or by adding or subtracting lights. Of course, using a waveform monitor is the best way to make sure you have even lighting. This will tell you immediately if you have contrast problems.

LIGHTING THE ENVIRONMENT

Now that you know the technical requirements for TV lighting, consider some specific methods for improving composition and mood of a presentation. There are two basic ways to light an

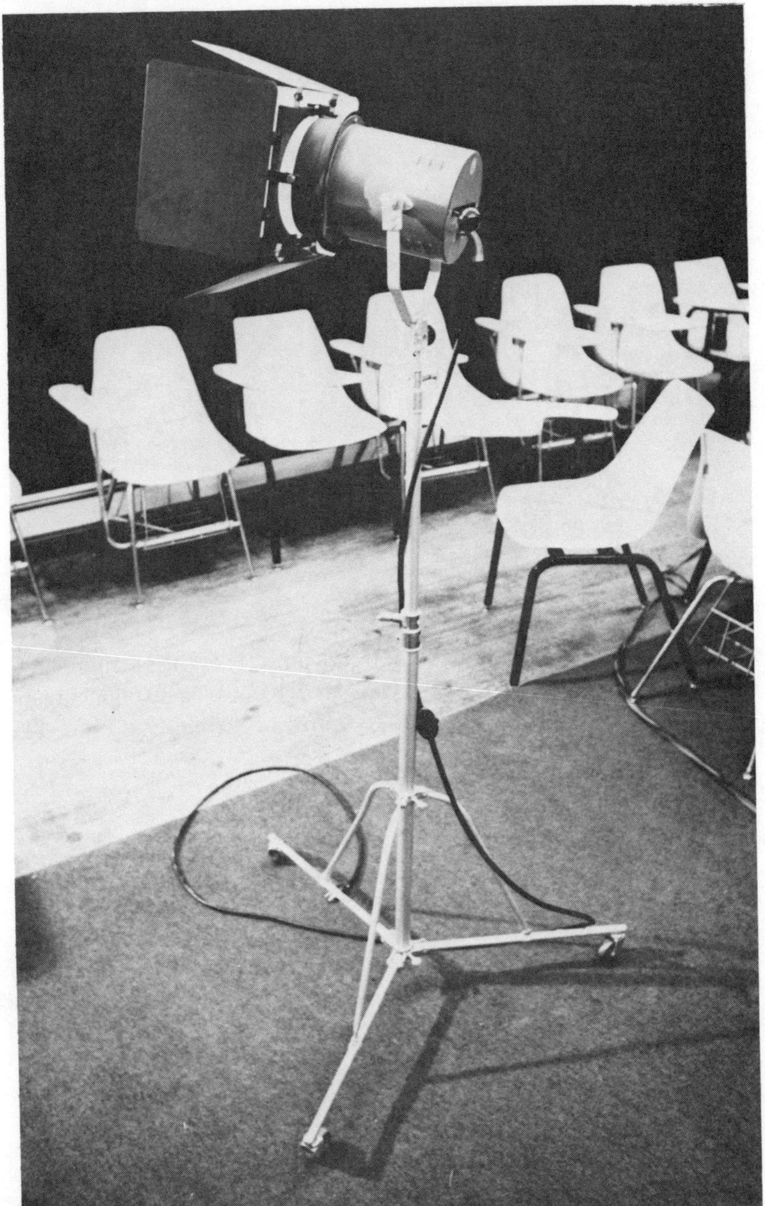

Fig. 4-3. A multibeam quartz light can be changed from a flood to a spot by moving the handle on the back of the fixture. The light is mounted on a stand and there are barn doors attached to it. Notice the on/off switch on the ac cord. This is a much better place for the on/off switch than on the fixture.

Fig. 4-4. A reflector is one of the most inexpensive yet useful light tools you can buy.

environment. First, you can point lights directly at the setting. This is called *direct lighting.* Second, you can light an area by bouncing or reflecting light on it. This is called *indirect lighting.* See Figs. 4-4 through 4-9.

DIRECT LIGHTING TECHNIQUES

When using direct lighting, I generally use two types of lighting instruments: the spotlight (or spot) and the floodlight (or flood). Spotlights generate a beam of concentrated light. Floodlights throw light over a wide area.

In looking at the environment to be lit, consider three areas—the foreground, the center of attraction, and the background. The most commonly used lighting technique is one called *three-point lighting.* It uses the following three lights.

Key Light. A key light is a spotlight set up directly behind the camera at a height that can vary from 6 to 8 feet. It is directed at the center of attraction.

Fill Light. A fill light is a floodlight set up to the side of the camera at a height of 4 to 8 feet. It is used to light the foreground.

Back Light. A back light is a spotlight set up behind the center of attraction. It is directed at the subject's head and shoulders.

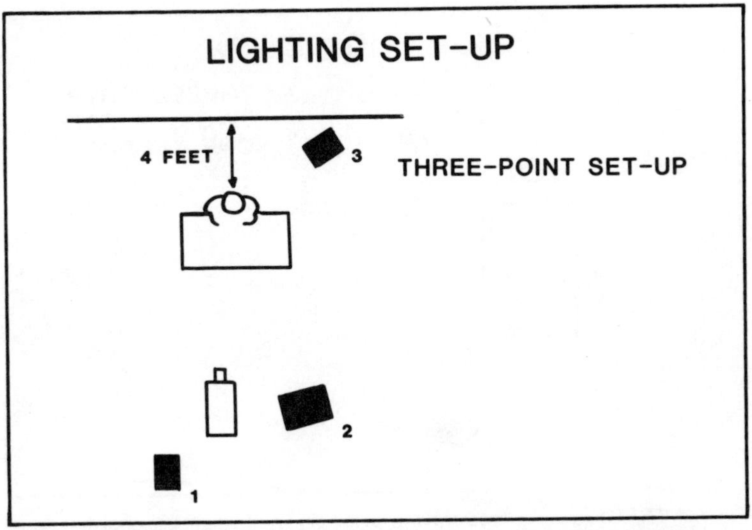

Fig. 4-5. A three-point lighting setup. The #1 key light is a spot set up behind the camera and several feet above it. The #2 light is the fill light (a floodlight) that is set up to the side of the camera and just a little higher than the camera. The #3 light is the back light that is a spot. It is set up to the back and side of the subject and as high as you can get it. Check your picture carefully to make sure you are not picking up the light stand being used for the back light.

Normally, you will find that the back light needs to be raised as high as possible. This is a very important light for TV composition. A TV camera reproduces a two-dimensional image. To give your pictures depth, you must create the illusion of a third dimension. The back

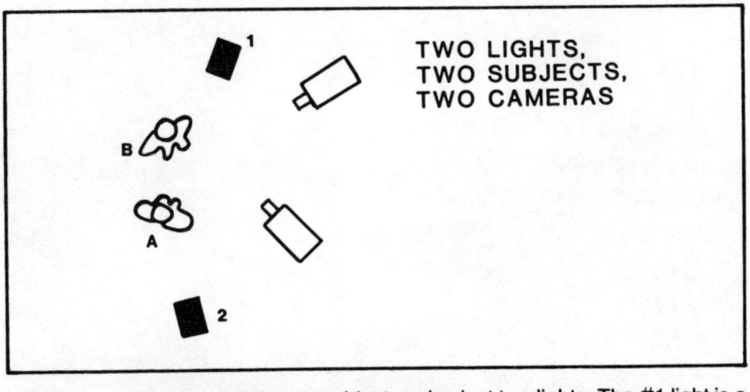

Fig. 4-6. A simple way to light two subjects using just two lights. The #1 light is a spot and serves as the back light for subject B and the key light for subject A. Light #2 is also a spot; it serves as the back light for subject A and the key light for subject B.

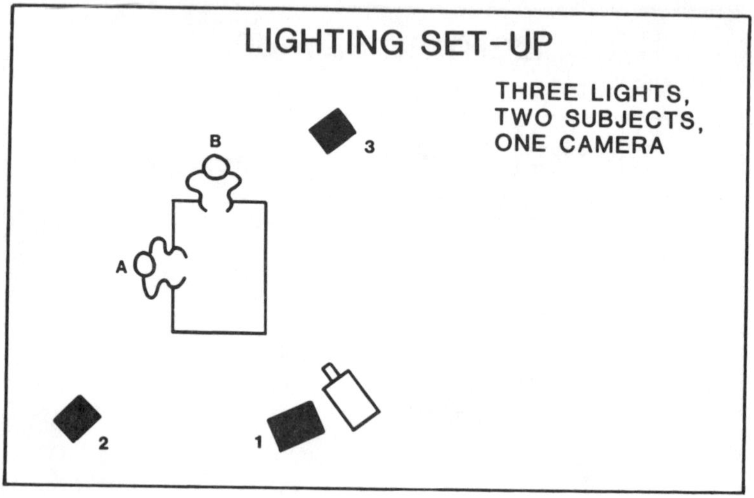

Fig. 4-7. The #1 light is a flood serving as the fill. The #2 light is a spot that is the key for subject B and the back light for subject A. The #3 light is also a spot that is the back for subject B and the key for subject A.

light can be used to separate the center of attraction from the background, this creates the illusion of depth.

Setting Up Three-Point Lighting

Use the following procedure for setting up a three-point lighting system.

Fig. 4-8. Looking at this one-light setup from the side. The camera does not see the harsh shadow because the subject is between the shadow and the camera.

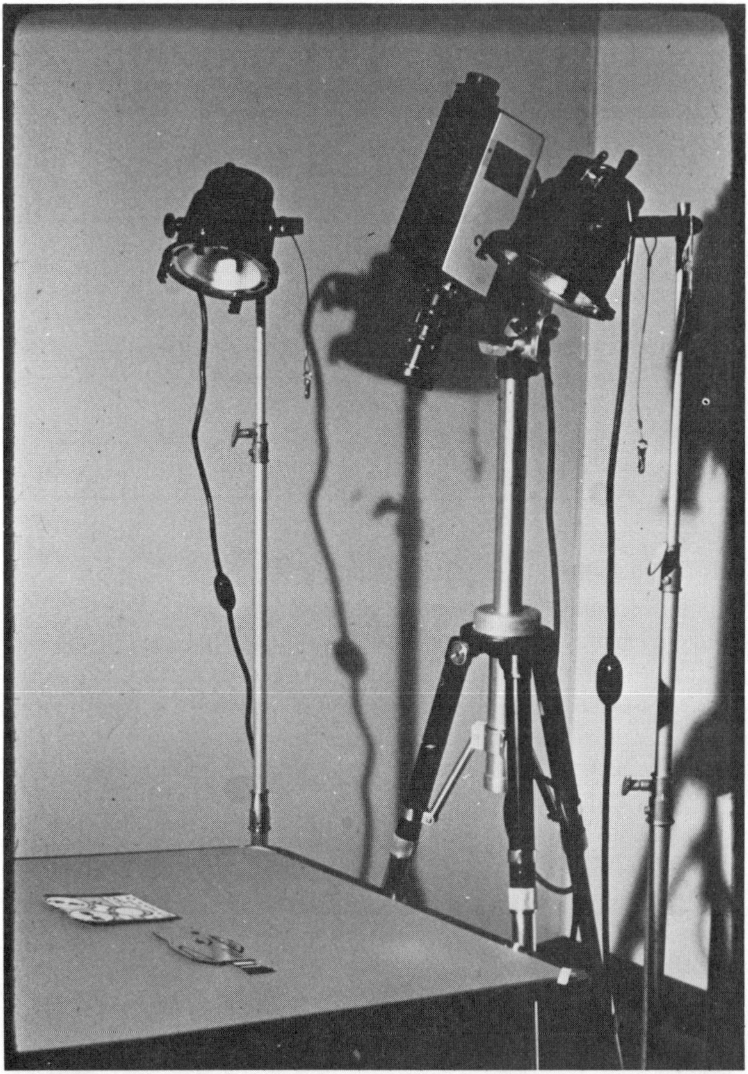

Fig. 4-9. A good setup for lighting graphics. Two 650-watt quartz units are mounted on stands at a 45-degree angle out from the camera. This produces good even light over the graphic. Mounting the lights at an equal angle reduces the possibility of unwanted reflections.

☐ Place your center of attraction at least 4 to 5 feet out from the wall. This helps eliminate shadows on the wall.

☐ Select your camera point-of-view (low angle, elevated, etc.).

☐ Set up your key light. After the key has been set, turn it off and set up the back light.

☐ With the key and back lights set, turn them both on and set up the fill light. The fill is used to light the foreground and to provide the environment with an evenly lit pattern.

In setting up lights, it is important to work with one light at a time. This is particularly true when you are setting up the key and back lights. This will minimize confusion when you are trying to correct shadows. Once lights are set, check your filter and do a white balance.

Once your environment is lit, you can experiment with extra lights. For instance, you might want to increase the light on the background. This can be done with a floodlight. Frequently, lighting the background will improve the picture quality because it might decrease the contrast ratio, and actually help define the center of attraction.

Setting Up One-Point and Two-Point Lighting

A *two-point* lighting system uses only the key and back lights. In an environment where you already have a fairly good fill, you will often need only two lights. If you are rushed or hurried, *one-point light* can be used—a single key light. This approach is used a lot in electronic news gathering (ENG) situations. A key or spotlight will generally be attached to the camera on a rod extending about a foot above the camera.

INDIRECT LIGHTING

If you must tape events in more than one area, you might want to use indirect, rather than direct, lighting. Let's say you're in a reception room. Some people are seated at tables and some are moving about, and standing and talking in groups. You should place a floodlight in each corner of the room and direct it at the ceiling. The light will bounce off the ceiling and raise the overall light level of the room. Before using this setup, however, consider these factors:

☐ What is the size of the room? What is the height of the ceiling? Four 1000-watt floods can be used in a room 30 × 30 feet. Larger areas might require additional lights on the sides, spaced equidistant from the corners. The lights should be 3 to 4 feet from the ceiling.

☐ The ceiling should be white. A dark ceiling will simply absorb the light.

One of the easiest and most flexible lighting tools to use is a good reflector (Fig. 4-4). It requires no power and it is an excellent tool for bringing up the overall lighting, (indoors and outdoors). You can make your own reflectors by using aluminum foil or even a white piece of cardboard. You can buy either silver or gold metallic reflectors. The gold ones give you a softer light. Use a reflector the next time you tape outdoors. You'll see immediately that it's a good and inexpensive way to improve your picture.

MOOD

Creating moods with lights is not easy in TV production. The problem is that the TV camera can handle a limited amount of contrast level. The following are some general guidelines for creating special effects with lighting.

☐ The use of a back light only will produce a silhouette effect.

☐ Diffused or softened light should be used to create the effect of sunlight. A window shadow pattern on a wall can also give the illusion of sunlight coming in from the outside.

☐ Different colored backgrounds can easily be created by lighting them with gels. A blue gel on a light directed on a wall can help create an illusion of "night" or "moonlight" coming in through a window.

☐ A low light or up light can create eerie shadows on a person's face and clothes.

Experimentation is the key to good lighting. You must first determine what your camera is capable of producing under different lighting conditions. Then, using the knowledge learned here, you can begin to upgrade the quality of your lighting standards. You'll often use a combination of direct and indirect techniques. Plan your lighting carefully *before* each taping session. Your reward will be superior tapes!

PRODUCTION TIPS

Here are some techniques to improve your lighting.

Shadows can be "hidden" by breaking them up with plants, bookshelves, or draperies.

If you are taping someone near a wall, sometimes there is a shadow on the wall behind him. If you can't move the person out from the wall, make sure your key light is mounted on a stand directly over your camera, and then lower your camera angle. This makes the shadow less visible.

Most people feel that one shadow is quite natural, but sometimes you might find that even a single shadow on the wall is too harsh. It can be softened by dumping a floodlight onto the wall.

When you are lighting an environment using quartz, you might find that your background is bluish. This is often caused by sunlight coming from the outside. There are two things you can do. First, you can try to block off the sunlight. Second, you can dump a flood on the wall with a red gel. The red light mixed with the blue light will neutralize the color.

Eyeglass reflections can be reduced by tipping the glasses forward. Watch for people with glasses that darken with more sunlight.

Shiny objects such as silverware will look better in reflected or diffused light. Dulling spray can be applied to tone down the objects' shiny surfaces.

To light graphics use the photocopy stand setup. Use two lights with equal strength set at camera heighth and at 45 degrees out to either side of the camera.

Whether taping inside or outside, always try to place your subjects in front of a backdrop that is relatively neutral in color, contrast, and composition.

You should have a basic makeup kit consisting of plain face powder and a roll of paper towels. Face powder is the best thing to use to tone down shiny spots or hotspots on peoples' heads.

Here are some rules to help you cope with "color problems" in your productions.

☐ Hard light yields brighter, more saturated color.

☐ Warm colors (red, orange, etc.) seem larger and nearer than cool colors (blue, green, etc.)

☐ Beware of reflecting colors.

☐ Small detail can cause colors to shift or disappear.

☐ A light color appears brighter against a black background.

☐ A dark color appears stronger against a black background.

☐ When several bright colors must be seen, use a neutral (18 percent gray) background.

☐ When choosing a color scheme for a set, refer to the color triangle and select any three colors that are harmonious. Using three colors is usually better than four; many times two colors will look better than three.

☐ Avoid white. Whites overload the camera circuits. Use a gray or a blue instead.

Clothes *do* make a difference. Keep these color tips in mind.

Blonds

Beige and/or dark blue are flattering.
Salmon looks fresh.
Avoid yellow because it can cast a violet reflection.

Brunettes

Medium gray and/or dark blue add intensity.
Light gray can give a tanned look.
Avoid yellow because it can cast a pink reflection.

Redheads

Medium gray and dark blue add intensity.
Light pinks are superb.
Avoid yellow through chartreuse colors.

Whites and Grays

Pink and rose are flattering.
Violet and blue can give pinkish fleshtones.
Medium and light gray can give pale, faded fleshtones.
Avoid bright, dark blue that can produce sallow fleshtones.

Chapter 5

Electronic Editing

There is no mystery involved in electronic editing. Like any precise mechanical process, editing proficiency requires training and practice. Don't be discouraged by technical difficulties. Remember that the editing process is only a means to an end. Mastery of editing is no guarantee that you can conceptualize, script, and produce worthwhile videotapes. Editing is an electromechanical process. Skill, timing, and practice will improve your editing, but only creativity, intelligence, and hard work will improve your tapes. Your goal should be to plan, setup, and execute edits that are so perfect—technically and logically—that your audience won't even notice the editing.

From the beginning, helical-scan editing was cursed by a lack of high-quality, reliable electronic editing. Broadcast TV and motion pictures have taught us to expect stability and continuity in our audio-visual productions. Except for "real-time" (straight through VCR production with its own attendant headaches), early "edited" videotape usually looked like it had been put together with a meat cleaver and Band-Aids (and, indeed, maybe it had!) See Fig. 5-1.

Video users soon discovered that attempts to make clean edits by stopping and starting the VCR or by splicing the tape (as in film or audio editing), produced disturbances in picture and sound. "Stop-start" editing and physical splicing difficulties are caused by the way in which the VCR places sound, picture, and control or synchroni-

Fig. 5-1. Before the development of electronic editing, programs had to be made in "real time." If a mistake occurred, the entire segment had to be taped over. This is an early 1-inch ampex helical scan VTR; my first VTR!

zation signals on the tape. It is easy to edit motion picture film because separate, succeeding frames are clearly visible. It's also easy to edit an audio tape because, with only one signal (sound), it is easy to locate an edit point.

Video tape carries three signals: audio, video, and control or synchronization. To further complicate matters, the video track is placed on the tape in a series of long, diagonal stripes; each pair forms a complete picture or frame. Stopping the VCR, placing it in record, and then starting it up again, erases not just one frame, but 20 to 30 frames. Cutting the tape vertically (as in splicing) interrupts several pairs of stripes and causes picture interference for many successive frames—before and after the edit point. On the other hand, the stripes are so long and so densely packed that it's almost impossible to splice them diagonally.

If you don't have an editing unit, you can break up the picture by going to "pause" between stopping and starting. You must work quickly because it's not a good idea to leave your VCR in the "pause" mode for more than a few minutes. The "glitch" that is produced can sometimes be camouflaged with a fade or some other type of transitional device (i.e., going out of focus, music, audio dub).

The only justification for physically splicing videotape is to salvage the usable portions of an otherwise irreplaceable production damaged during playback. In all other cases, you'll save time, trouble, and apologies by avoiding physical edits. One additional note is that if you *must* splice videotape, beg, borrow, or buy a videotape splicer (an investment of about $20.00). Use a recommended splicing material. Don't do it freehand!

In 1971, a new kind of editing evolved. Following the pattern in the 1-inch VTRs, some ½-inch, open reel machines were offered with "Capstan-servo" editing, featuring "flying" eraser heads. Oversimplified, this means that sensing circuits allow the video recorder to do electronically what you are unable to do physically. The machine lays down both a new "field" and an old "field" of video information.

So we finally had an editing VCR that created clean edits. It worked fine when the source was a live camera. When the source was another prerecorded tape, you had the problem of making sure the two tapes—the source tape on the playback unit and the edited tape on the editor—arrived at the appropriate points simultaneously. You just can't start up to VCRs and push an edit button!

When the source material is being originated from a prerecorded videotape, you must first make sure that the signal is up to speed. In other words, you must make sure that the source VCR is providing the editing VCR with a proper sync and video signal. To get this, you must make sure the source VCR is playing back at normal speed.

An essential factor in assuring clean edits when editing tape-to-tape or machine-to-machine was having both the source and editor play back at normal speed. From a "stop" position, this required 10-15 seconds from the time you pushed the play button; from a "pause" position it took 3-5 seconds.

In the mid-1970s, video/computer technology helped us out of this ulcer-producing predicament. Small units called *backspacers* or *editing control units* (ECU) were developed. The ECU is connected to both the source and the editor. It controls the play and record functions of the editor and the play function of the source unit.

The ECU works this way. First, you find the point on source tape where the information you want to add to the editing tape begins. Next, press a cue button. The ECU then stops the source tape and backs it up 5 seconds. The control or sync track is used to measure out the time. This is why this type of editing is sometimes called *control track editing.* Next, find the point on your editing tape where you want to add the material. When you press the cue button, the edit tape is backed up 5 seconds. As you press the "perform edit" button, both units start and run in playback for 5 seconds. The editor performs the edit and the job is done!

Today we have ECUs that come in all shapes and configurations. Depending on the cost, an ECU can offer a number of control functions such as the following.

- ☐ Tape search in both forward and reverse at speeds from one frame at a time to 2-to-3 times normal speed.
- ☐ Preview edit is a function that allows you to preview your edit before it is performed.
- ☐ Edit problem alarm or alert devices that will tell you, before you begin an edit, that you have a problem with your video or sync.
- ☐ Frame and second counters that give you an LED readout of frames, seconds, and minutes.

The more sophisticated editing and ECU systems today no longer use the control track for "timing" purposes, they use a *time code* instead. This is a computer signal that is produced by a time code generator and recorded on one of the audio tracks or on a separate time code track. The time code is fed into the ECU to provide the user with a digital readout that he can use to log and edit the tape as well as program the ECU.

EDITING SETUPS

In setting up an editing system, I refer to the piece or pieces of equipment supplying the audio and video information to be edited as the *source gear.* Here are a number of basic setups that can be used to generate a live as well as prerecorded source signal. See Figs. 5-2 through 5-5.

EDITING MODES

Editing VCRs can perform two basic types of edits: an *assemble edit* or an *insert edit.* Use of the assemble edit mode results in an edit where both audio and video information are added in serial sequence (A, B, C, D, E, etc.). Use of the insert edit mode gives you a number of options. You can edit video or audio only in a serial sequence (A, B, C, D, etc.). Or you can insert audio and video, video only, or audio into an existing sequence at any point.

Because it is more flexible, the insert edit mode is the most popular. The only requirement for using the insert edit mode is that the tape you are inserting on must have a control track or a sync track. This can be accomplished in one of two ways. First, you can

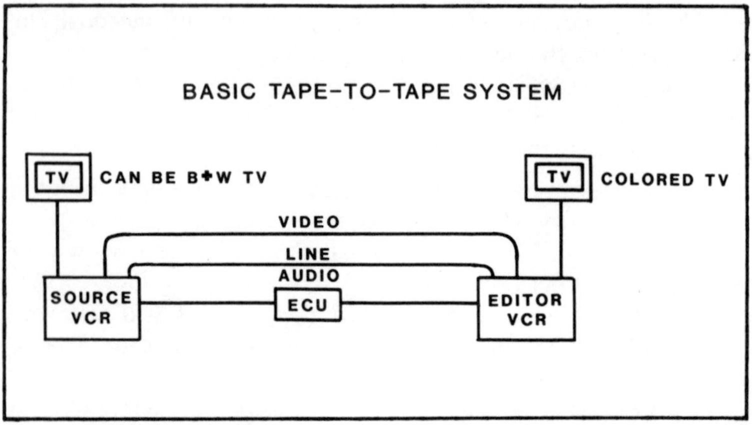

Fig. 5-2. Editing system setup.

feed the camera signal (camera lens closed) into your VCR and record it onto a new cassette or, second, you can use a tape that already has information recorded on it. This sync track provides a lock-up signal for the incoming signal. The sync track should be laid down in an uninterrupted sequence. If a break occurs in this sync track, it will interrupt the lock-up of the incoming signal and usually produce a glitch, picture roll, or break-up at that point on the edited tape.

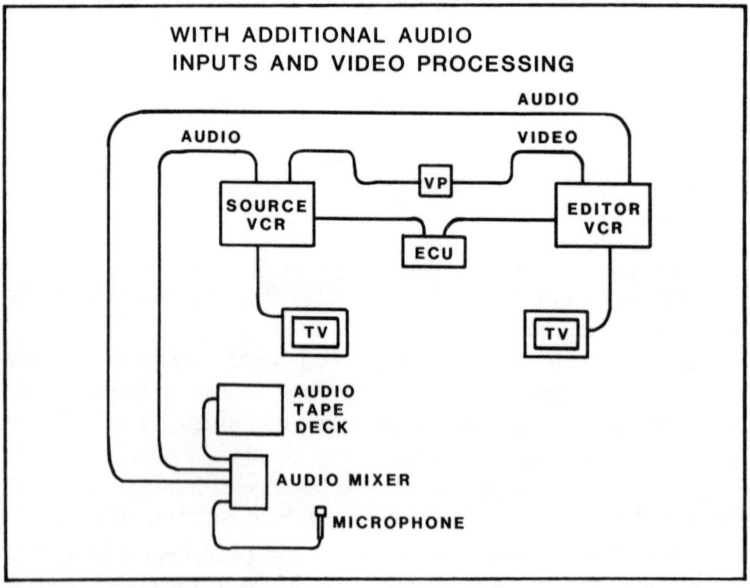

Fig. 5-3. Editing system setup.

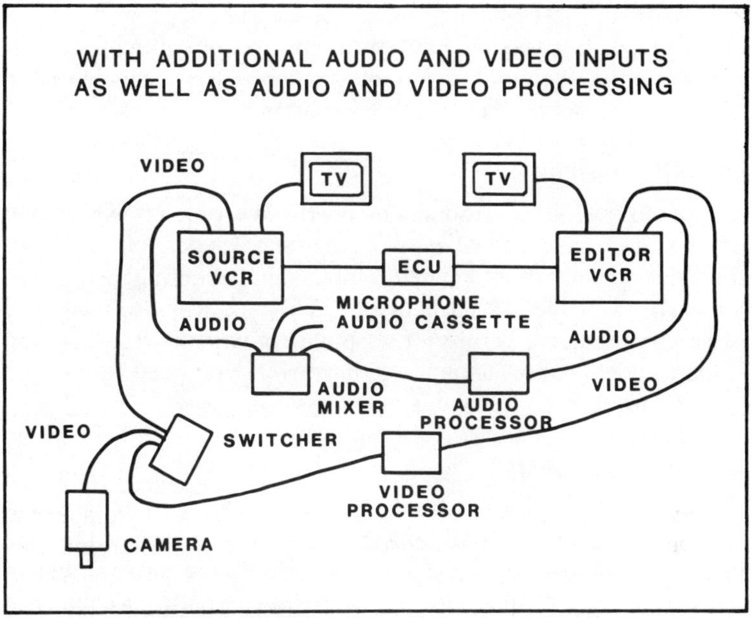

Fig. 5-4. Editing system setup.

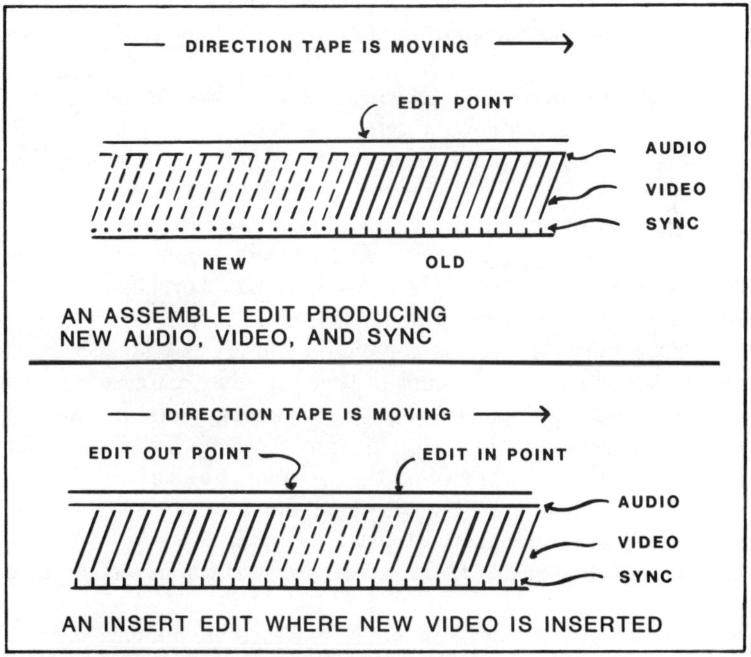

Fig. 5-5. How assembly and insert edits are produced.

Using the assemble edit mode does not require the presynced editing tape. But you are limited to one type of editing—straight assembling of audio and video information. Also, assemble edits generally are not as technically good or as clean as insert edits.

EDITING STRATEGIES

Almost all video productions need some type of audio/video editing. How much and what kind of editing depends, of course, on the message you want to convey and the kind of editing equipment you have available. Described here are some ways to plan your programs—under a variety of shooting conditions—so when you come to post-production you will have what you need to create a finished product that conveys your message.

NO-SCRIPT PROGRAMS

Let's begin with the least complicated of all shooting situations: one camera and one cameraperson. The subject of the tape might be a demonstration, a lecture, a performance, or any other ad lib content area. In such cases, the camera is usually portable. You go to the subject rather than bringing the subject into your home or studio. Needless to say, such productions are done in real time and without rehearsal. You get it the first time through (in many cases) or not at all.

Your first concern is to get all of the essential information recorded on the tape—both video and audio. Depending on the circumstances, this might mean planting yourself in one position and staying there. If this is the case, you must depend on your zoom lens to catch close-ups and provide some visual interest. Under other conditions, you might be able to walk around and provide interest by shifting point of view. Always remember, however, that what you see is what you get.

Once the basic tape is completed you are not finished. You should spend an equal amount of time on other material such as details, close-ups, reaction shots, establishing shots, and similar content. These do not have to come from the same situation as your basic program. For instance, if the program is a one-time lecture and you want shots of the listeners reacting to what is said, you can shoot any group of people reacting as long as the reactions match the content of the speech. If you are pressed for time, you can tape stills.

Let me give you an example from a tape we did during a folk festival on the Mall in Washington D.C. We wanted to document the varieties of music being performed so we taped several groups through one or two numbers. We kept the camera in one position and focused (in most instances) on medium or full shots of the performers. Once that basic tape was completed, the cameraperson spent time taking additional shots of the event from various angles. Included were long shots of the group, shots of individual faces of people in the audience, shots of the audience as a whole clapping in time to the music, wide views of the Mall with people walking by, etc.

In the studio, log and index the entire tape (both the basic information and all of the extra material). You have to be able to find that close-up of the freckled kid with the big grin or the one of the old man's foot tapping in time to the music.

The first editing job is to condense the primary content tape so that it will include all of the essential information, but omit extraneous material. In "ad lib" situations, there will almost always be pauses, repeats, asides, out-of-focus camera shots, and other information that is not important and should be omitted. Once you have identified the essential segments, edit these together in their proper order.

Now go through the assembled primary tape and identify where you need inserts to cover up bad shots or *jumpcuts* (abrupt shifts in the action created during the earlier assembling process). This is where you will use the cutaways you shot after documenting the primary event. The cutaways (usually video only), when inserted, will provide you with transitional devices that can greatly aid in strengthening the overall flow or continuity of your program.

DOUBLE SHOOT

If you have a larger staff, you can do no-script programs with two people, each with his or her own portable system. In this case, one camera photographs the speaker(s) while the second takes pictures of the audience, shoots the speaker(s) from different angles, etc. See Fig. 5-6.

It is desirable, when possible, to feed both recorders the same source of audio. You can do this by using a mixer or amplifier. This will allow you to avoid problems in the editing process with lip sync and noticeable differences caused by the contrast between the audio produced from a mic on the speaker's desk or podium and the audio from the mic built into the camera that is moving around the room.

In the studio, again log and index both primary and secondary tapes. If the presentation is lengthy and needs to be condensed, it's

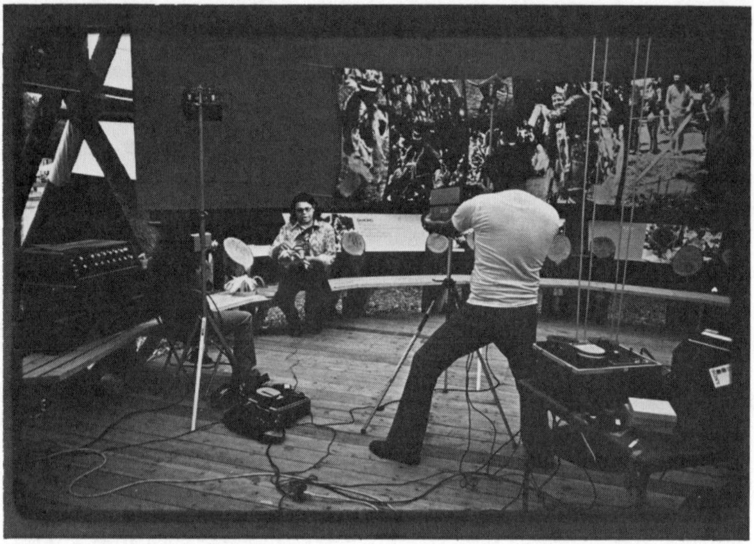

Fig. 5-6. This event is being double shot by two separate single camera systems. One person covers the subject with medium and wide shots while the other camera person covers the action with close-ups.

easier to make a transcript than to view tapes over and over. Use the transcript to make the needed condensations and provide the desired continuity. As before, assemble the basic content from the primary tape and insert segments from the secondary tape to provide visual interest and add rhythm and emphasis to the program.

SCRIPTING IN THE FIELD

If you must do a documentary in the field, you might need to develop a script as you go. I faced this problem when I did documentary tapes on Indian activities for the Bureau of Indian Affairs. In this instance, a series of interviews with various leaders and members of the individual tribes formed the basic tape content. The interviews were shot first. That evening I watched what I had done during the day and made a list of the cutaways that I wanted to illustrate the narrative in the interview.

The next day I returned to the site and shot cutaway material. Again, shoot more than you think you will need. If you think you can use 10 seconds of students doing a handcraft in a classroom, shoot at least a minute. Better still, shoot several minutes from different perspectives and angles. This provides a choice so that you can use that one particular shot that best illustrates the content. Also, try to get audio to go along with the video. You might find in editing the tape that some of the cutaways can be used as transitions. If they have their natural sound, it will add to your production's style and quality. Make a transcript of the audio content (in this case the conversation from the interviews). This becomes the transcript.

In the studio, the editing process is the one previously discussed. Put together the primary source tapes (the interviews) along with their audio and edit in the cutaways. Mix in audio under the basic narration or audio content as appropriate.

FULLY SCRIPTED PROGRAMS

Editing together a program from a written script is probably the most efficient means of producing a tape. Make sure the script is written by someone who understands your editing capabilities and limitations. Once the script is written, you must decide on the visual or video actions that will accompany the narration. This information can be added to the script using line drawings, quick cartoons, Polaroids, or short written descriptions.

Once the script is prepared, draw up a shot list. The shot list outlines the taping sequence. It does not have to be the same sequence called for in the script. In the post-production process, you can edit the "scenes" together in the proper order. All scenes should be slated (i.e., cassette #1, scene 2, take 1, etc.), and all cassettes and boxes should be labeled properly. Catalog and index the scenes when the shooting is completed. I put all my source material on cassettes (three to four scenes to a cassette). I find this saves me a great deal of fast forwarding and rewinding as I search for a certain scene or take in the editing process. Of course, the source tapes can be reused once the production is complete.

The editing strategies discussed so far have all used tape-to-tape editing systems. Several strategies can be used for a "live" editing system (camera or cameras to editing deck).

EDITING ON THE RUN

Most portable units today have an assemble edit capability that allows you to produce programs by editing "in the camera." This technique is frequently used in production of 8mm home movies. Producing programs this way is difficult because there is no second chance. What you shoot and where you edit is what you get. To use this method effectively, you have to think ahead. Before taping the event (interview, demonstration, etc.), think about the transition shots you will need. Decide on the length of the overall program and schedule your taping of the "main events" and transitional shots accordingly.

There is one post-production technique you can use here called the audio dub. This allows you to place either new narration or music over prerecorded video segments.

Most decks with the assemble edit capability will perform the edit from a pause position only. This means that, after you tape a segment you can:

☐ Put the tape into pause and move on to your next event (but remember you shouldn't leave the tape in pause for more than a few minutes).

☐ You can rewind the tape and check out what you've just taped, then run the tape up to the point you want to edit, put the unit in pause, and continue your program.

Editing on the run is a good strategy for producing short, ad-lib tapes. It's also effective in situations that can be easily controlled and structured. Examples are the "house tour" tapes made for real estate salespeople. See Fig. 5-7. Additional applications for editing on the run are discussed in Chapter 10.

LIVE-ACTION EDITING

Live-action editing is often used in making short training or

Fig. 5-7. Editing on the run means having all your gear ready and accessible. Know what it is you want before you start to shoot.

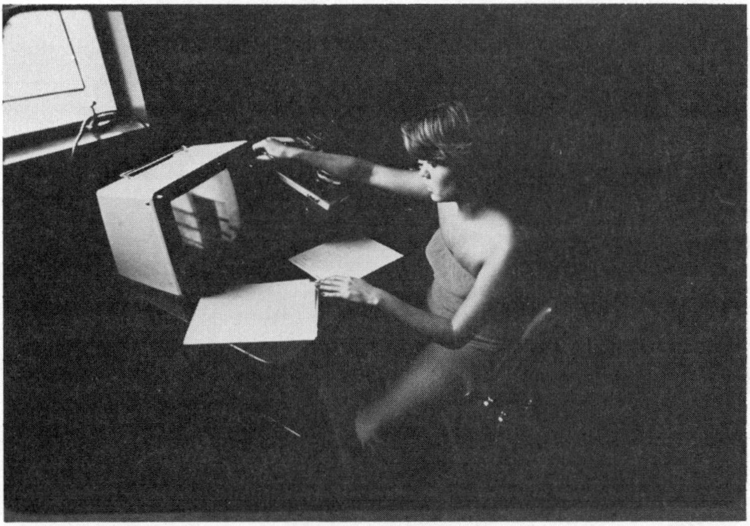

Fig. 5-8. Careful logging and indexing of your source material will make the editing process go faster and it will be much more enjoyable! This person is logging source material using the log sheet form shown in Fig. 5-9.

informational programs (especially those involving "how-to" instruction). First, the narration is placed on the tape using the audio insert mode. The video or "live-action" portion is then inserted to match or supplement the audio. Graphics consisting of diagrams or titles can be inserted along with the live action.

Once the narration and video have been recorded, a music track can be laid down on the opposite or unused audio channel. By placing your VCR's audio mode outlet control in the "mix" position, you can listen to the mix as you audio dub the music. This way you can properly adjust the level of the music so that it won't override the narration. Make sure, when the tape is played back, that you place the unit in "mix" and that you run your audio out of the monitor line output.

LOGGING AND INDEXING THE TAPES

The editing process is a lot more enjoyable when you know exactly what and where to find the scenes to be edited. This requires a good system for logging and editing your source tapes (Figs. 5-8 and 5-9). Here are some guidelines.

- ☐ While in the field, label, number, and date all source tapes.
- ☐ Use an accurate means of indexing the location of scenes.

PROJECT Lower Sioux TAPE NO. 001-3 DATE 8-76
O ______________________________ OF TAPING
⊙ ZERO COUNTER AT AUDIO CUE

COUNTER NO.	AREA	DESCRIPTION	COMMENTS
005	F/FB	Personal information is related	Name did not come through on sound track
087	H/C	Community information and history is related	
179		Break	
185	H/C	Outbreak of 1862	
250	G/L	Tribal Council programs are discussed	
297		Reorganization Act and its effects	
300		Discussion of other council programs	
341	A/C	Information about personal interest (painting) and family painting history is related	Name not clear; appears to be "Ruben"
357		Art training (personal)	
383		Professional art career is discussed	
423		Works of art are displayed	
442		More information about professional art career is related	
454		Hunting is another personal interest	
470		Teaching high school art	
485		Introduction to Sylvia Spencer who describes her role as assistant	
503		Break	
504	H/C	View of artifacts	
552	S	Example of Sioux language	

Fig. 5-9. Log sheet form.

Use your VCR counter for this, and especially if it counts either frames or minutes and seconds. You can still use your counter if it counts "reel" rotations, but it won't be as accurate. Also, "reel" counters vary from one VCR to another.

☐ Keep your log simple. Note any technical difficulties such as bad audio or interference in the video. Avoid making "esthetic" comments at this point because, until you know the context in which the scene will be used, you really don't know if the scene is usable or not.

ANIMATION

The newer ECUs (editing control units) and editing VCRs are very stable and provide accurate edits within a frame or two. This capability permits you to do some animation for your graphics by using pop-ons and other simple animation techniques.

When changing the graphic, be sure not to move it. Also, be

sure that the camera does not move. Even the slightest shift in angle or distance will be very noticeable in the finished program.

Lay the audio track down first. Then insert your graphics. The closer and more consistent the edits, the smoother the animation effect will be.

VIDEO DUPLICATION

Electronic editing is simply a form of selective duplication. The basic difference between an editing system and a duplication system is that the latter requires regular play and record VCRs—not a special editing VCR. You also need only one TV to monitor the output of the "copy" unit. There are many ways of setting up an editing or duplication system. Above all, remember to keep your video signals as clean as possible. In other words, don't decrease their strength or impedance by using improper cable, connectors, or splitters. Make sure your gear is properly terminated.

PRODUCTION TIPS

When shooting tape in the field, always leave 10 to 15 seconds of preroll time on the front of all takes. This allows your editing unit time to lock-up with the oncoming signal. I place a "slate" or card on the camera during this preroll time to identify the particular scene and take.

When using the insert edit mode, be sure that the tape on your edit VCR has an uninterrupted sync track. This can be created by placing a tape on your VCR, feeding a faded down camera signal into it, and recording the signal either for the entire length of the tape or at least for 10 to 15 minutes longer than you anticipate your edited program will be. If you find that you must go beyond this point, stop your editing, hook up your camera, and add more video sync by doing an assemble edit.

Before doing an edit, check your tracking, skew, video level, and sync level.

Tape is inexpensive (you can use it over) so when you are on location shoot plenty of footage.

Label all source tapes and boxes. Always check each take. This doesn't mean checking the entire tape. Usually, if the last 60 seconds is OK (audio and video), the rest of the take will be OK too.

In live action editing, once the narration track has been laid down you can use the unused audio track for camera or action directions or cues such as fade ups and downs, hand movements, etc. Once the tape is completed, this track can be erased and a music track can be laid down.

The key to good copies or good edited masters is a good source signal.

Once you have set up your system, check all cables and connectors to make sure they are properly tightened and secured.

Always make a test recording to check system operation.

Check each edit.

As with anything else, the most important factor in good editing is practice, practice, practice. There are no short cuts to developing the feel for the flow and dynamics of an effective program that good editing can provide. But such practice need not be "work." There are many ways to have fun with video editing. Here is one idea.

First, select a piece of music you like—anything from Vivaldi to Jethro Tull. Take your portable equipment outside and shoot a street scene: the faces, hands, and feet of the people passing by; crowd shots; busy intersections; people going in and out of stores, etc. Back in the studio, lay the music down on the audio track, and then edit the video to complement the dynamics and continuity of the music. Play around and experiment. It's all good practice and experience.

Don't try to do more than one thing at a time. If you are working with the audio, think audio; don't worry about the still photos you intend to use to illustrate certain points. If you are dealing with graphics, think graphics, and so on. Trying to do everything at once often means that nothing will be done as professionally as possible, and something will be forgotten entirely. After the program is in the can is *not* the time to remember that you intended to use footstep sounds during that shot of the boss walking down the hall to his office.

Remember that good edits are not seen. You want the viewers to concentrate on the program content (not on the production techniques used to present the content). If the editing is inept, it calls attention to itself, and the audience becomes conscious of the way the tape is put together instead of what the program is saying.

Always keep the total concept of the program in mind, plan ahead, and shoot more than you will need or use. Tape, after all, can be re-recorded. If you don't get it the first time, you might not have another chance. Editing *is* for everyone. Isn't it time you took the plunge?

EDITING TIPS

Here are a few important factors to keep in mind when undertaking editing.

The key to good edits is good quality original or source material. Remember, any problem in your source or original will not only be magnified in your edited master, it will also be more apt to cause instability in the edits themselves. It will certainly show up in the duplicates.

Always check all connectors and cables in your editing setup. Make sure, for instance, if you are using looping monitors that they are properly terminated.

Clean audio and control track heads as well as all tape guides before each editing session. Clean the capstan, etc.

When originating the source tape, try, when possible, to use the best sync available. This will significantly affect the quality of edits.

Always check audio and video levels before each edit.

Use a waveform monitor to monitor sync and video levels during editing.

Play back each edit to make sure that you achieved the best possible quality your VTR is capable of giving you.

Evaluate your equipment requirements in terms of your shot list. Plan carefully for the equipment and crew.

In your interviews, studio recording, and location shots, always record 6 to 10 seconds of stable tape before and after each take.

Check each source take immediately after recording.

Before you begin the actual production process, catalog and index all source tapes.

Make an editing script outline to coordinate and structure all of your material.

Strive for the best possible audio and video quality on your *source* tapes. It will pay off in the editing process.

Chapter 6

Multicamera Systems

There are certain disadvantages to using the multicamera model:

- ☐ It doubles the necessity for good solid planning.
- ☐ It adds to staff requirements.
- ☐ It adds to the human/equipment unreliability quotient.
- ☐ You are stuck with the shots you get.

On the other hand, there are many advantages:

☐ The multicamera/SEG (special effects generator) system offers the user a variety of effects not obtainable with a single-camera editing system.

☐ It allows shorter turnaround in the documentation of ongoing events. This is of particular importance to CATV because low production budgets and tight program scheduling don't permit time for effective editing.

☐ The multicamera/SEG system gives a single individual maximum control when covering an on-going event with a variety of camera views.

☐ It's more difficult and time consuming to resimulate real time flow when using electronic editing.

It's not true that the multicamera system is being replaced by editing. The most flexible video system is one that incorporates a multicamera/SEG and editing system. The basic components of the multicamera/SEG system includes an SEG, a viewfinder camera, a camera preview monitor, and a program monitor.

The SEG. Although you can effectively operate a multicamera system with a regular video switcher, most multicamera production systems incorporate the SEG for added flexibility (fades, supers, etc.) For an SEG to properly handle all the incoming camera signals, it must supply each camera with a common sync signal. Most SEGs have a built-in random sync generator. They can also accept the sync drive from a more sophisticated and superior 2:1 perfect interlace sync generator. The signal from such a unit is then used as the common sync to drive all cameras.

Viewfinder Camera. One advantage of a multicamera system is that it gives highly controlled multicamera coverage of an event. To maximize this situation, the camera operators must see what their cameras are seeing. Thus the need for viewfinder cameras.

Camera Preview Monitor. Because the key to effective multicamera operation is the director's ability to coordinate camera movement, he must be able to communicate with the camera operator and see what the cameras are seeing. Only then can he direct the camera operator and make sure that the proper picture is on the camera before he switches it into his program line. To do this, each camera should have its own "preview" monitor. Some of the more expensive SEGs also have an effects preview monitor that enables the director to check out certain special effects before switching them on line.

Program Monitor. In addition to seeing what the camera operators see, the director must also see what is coming out of the SEG into the system or into his VTR/VCR. The program monitor can be hooked up just out of the SEG or from the VTR/VCR.

Don't underestimate the value of good director/crew communications. Don't think that multicamera coverage will alleviate the burden of decision making associated with electronic editing. The person operating the SEG is involved in a high degree of decision making that requires making decisions in "real time." A lack of an adequate communication system between the director and his camera crew affects the director's ability to make split-second changes in his production strategy (always a necessity when documenting a real time ongoing event).

MULTICAMERA HOOKUP

As with a single-camera system, a multicamera system should be hooked up one pathway of information at a time. With the single/camera system, you must usually deal with three basic pathways of information: audio, ac power, and video. With a multicamera system, you have two additional pathways: the sync and the intercom system. Before any gear is connected, it's a good idea to make sure that each individual piece is in working order—particularly the cameras, monitors, and VTR or VCR. The SEG is difficult to check without a scope, but checking all your gear carefully saves time and effort later on. See Figs. 6-1 through 6-4.

Table 6-1 describes some problems you might encounter when hooking up your multicamera system. These are problems that beginners as well as experienced professionals encounter. Keep in mind that by adding cameras, monitors, and microphones, you multiply the number of things that can go wrong. When a problem develops, check it out systematically. Trace the appropriate pathway (ac, video, audio, sync, intercom). With a multicamera system, this is the only way to check out the system and remain reasonably calm.

MATCHING CAMERAS

When switching between two or more cameras, be sure the colored image reproduced by each camera is the same (camera one reproduces the same red hue as camera two, etc.). Using cameras of the same model number made by the same manufacturer makes this task a lot easier. Also, cameras with external chroma, burst, and

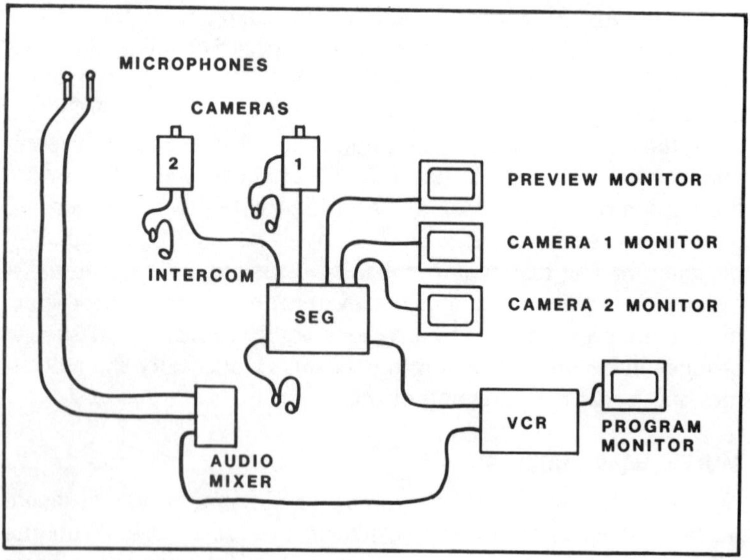

Fig. 6-1. A multicamera system setup.

Table 6-1. Multicamera System Hookup Problems.

Description of Problem	Cause	Solution
Picture wiggles, rolls is overall very unstable.	No sync	Sync cable is not properly connected. Sync generator or SEG is off. The internal/external switch on the camera or the SEG is in the wrong position. Bad sync cable.
Picture is doubling or ghosting	Unterminated connectors	Check connectors on all preview and program monitors and looping output of your SEG.
In fading from one camera to another, you get noise or black streaks in the picture.	Your fading controls are dirty.	Spray with freon or other appropriate cleaner recommended by dealer.
No picture in camera preview monitor.	SEG is off. VTR/VCR is not in record. SEG is in the wrong effects mode.	Make appropriate changes in settings on gear.
No picture in preview monitor.	Camera is not on. Loss of or defective video cable.	Make appropriate changes in settings on gear.

Fig. 6-2. A typical multicamera setup with three camera preview monitors and a program monitor.

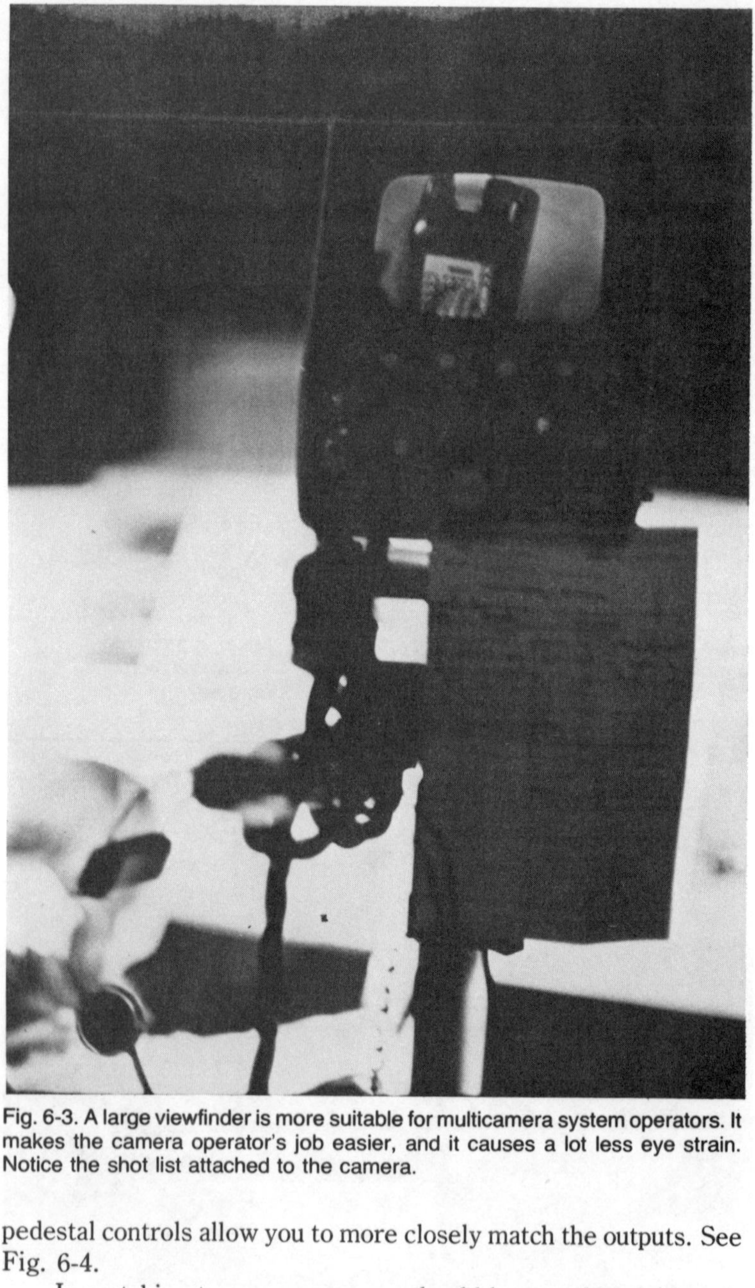

Fig. 6-3. A large viewfinder is more suitable for multicamera system operators. It makes the camera operator's job easier, and it causes a lot less eye strain. Notice the shot list attached to the camera.

pedestal controls allow you to more closely match the outputs. See Fig. 6-4.

In matching two cameras, you should have a chart with red, blue, and green colored chips or strips. You can also use a bright red, blue, and green poster. Set up your chart or poster and light it with the lights you'll use for the taping. Focus your two cameras on the chart ; make sure you get the same picture with each camera. Check your filters and set up a white balance. Now switch back and forth between the cameras. If you get a distinctive color change or shift, you probably have phasing problems. Normally, this can be adjusted with a control in the switcher. If you have no external controls (chroma, pedestal, etc.), about the only thing you can do is check the iris openings. After you have matched your cameras using the chart, check them out for skin tones. You probably won't get perfectly matched cameras, so select the camera that best reproduces skin tone as your close-up camera and use the other for cover shots.

SEG OPERATION

Having reviewed the setup and hookup, you can now turn your attention to the effective operation of the multicamera system. This involves knowing some basic operational procedures. If you have never operated a multicamera system before, and you have an SEG, first "play" with the switcher—switching from one camera to the other. Then move on to another effect such as fading. Take one effect at a time. Practice with it and then produce a short tape using just that effect.

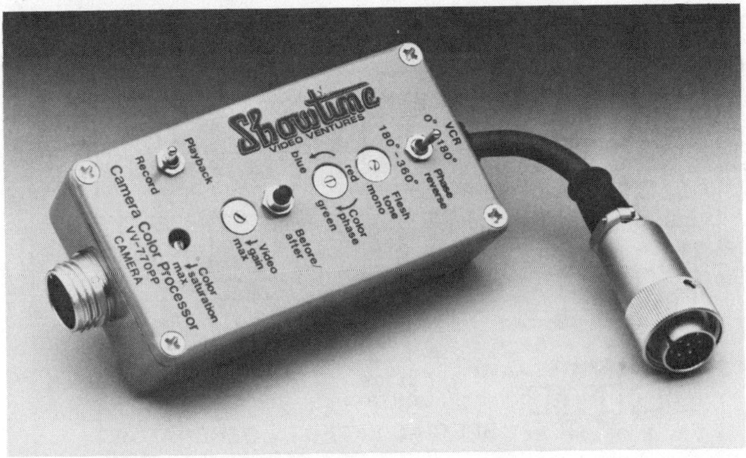

Fig. 6-4. This unit can be placed in line with a camera to aid in phasing or matching it to a second camera in a multicamera setup. It can also be used to control and correct variations in tapes being played back for duplication or editing (courtesy of Showtime Video Ventures).

You can learn to operate an SEG in a few hours or a few weeks. Much depends on the complexity of the unit. Learning to operate it is just half of learning to use it. Practice is the key. Start out with a two camera setup and then work in a third or a fourth camera. One of the biggest problems in using an SEG in a production situation is going from one effect to another. To do this skillfully and quickly requires a good understanding of how your SEG operates.

Take a moment to study the diagram of an SEG shown in Fig. 6-5 and the explanation of how it works. The row of buttons marked PROGRAM determine what signal will be sent out of the SEG into your VCR. You can switch back and forth between cameras 1 through 4 by pressing the appropriate button. When the SE (special effects) button is depressed, you must also depress either the dissolve or wipe button. If you press the dissolve button, you can dissolve between two signals, using the slide control. Select the two signals you want to dissolve between by pressing the appropriate buttons on the A and B bus rows. If you use the WIPE mode, you must select a particular wipe by pressing one of the wipe effects buttons. The slide control is now used to adjust the wipe effect. The two signals used in the wipe are determined by the buttons pressed on the A and B bus rows.

The PREVIEW buttons when depressed will display either a camera or a special effect on your preview monitor only. It will not send the signal to your VCR. This preview feature lets you set up

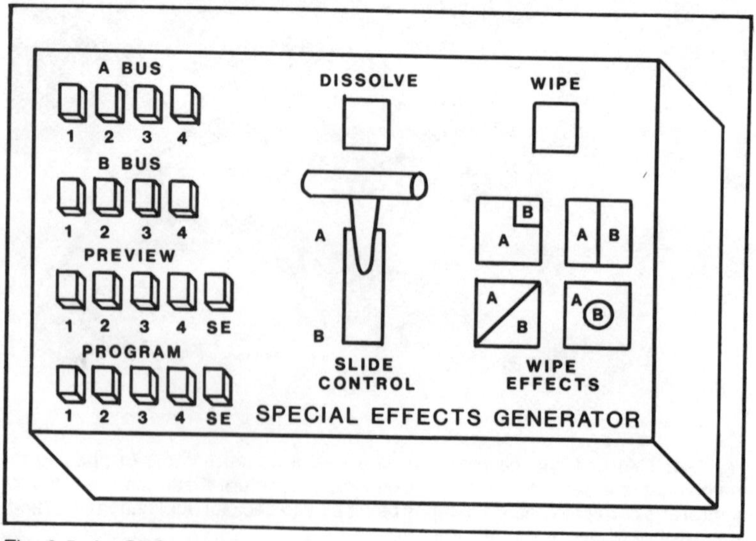

Fig. 6-5. An SEG diagram.

special effects shots. For instance, to set up a corner insert effect you could do it this way:

☐ Focus camera 1 on the subject. Send camera 1 out to the VCR by pressing "1" on the program row.

☐ Press the SE button on the preview row.

☐ Press camera 1 on the A bus and camera 2 on the B bus.

☐ Press the wipe button and the camera insert wipe effects button.

☐ Use the slide control to set up the corner insert. Have camera 2 frame the person on the camera insert.

☐ Now press the SE button on the program row. You'll be feeding out the corner insert shot to your VCR.

To assist you in setting up effects like split screen, you might want to use a nonpermanent marker and actually draw a line on the face of each camera's viewfinder to indicate the areas that camera is to cover. Make sure these lines correspond with similar markings on your program monitor.

One secret of effective camera movement is having a reason to move the camera. And one secret of the effective application of the SEG is having a reason for every special effect. Remember that mixing effects sometimes confuses the viewer. This is true even when choosing between switching or fading. Spend some time thinking about what you want to do before you do it. This is called preproduction planning, and it comes wrapped in a box with a clock.

At first, the SEG with all its buttons and effects will boggle even the best of minds. Nevertheless, it can be exciting and rewarding in the coverage of real time events *if* you do your preproduction planning, and if you know what's going to happen ahead of time. Avoid using the SEG to document ongoing real time events that are going straight out over the air or events where you have no idea what's going to happen. It takes an experienced director to handle this type of situation. Don't take it on until you've done some practicing.

If you do get involved in such a situation, the best way to cover it is to give your camera operators specific duties. For instance, have one camera "follow the action," using medium and close-up shots. Have a second camera "cover the action" using a wide shot. The second camera's assignment is very important. It's always ready with a cover shot when the first camera can't follow the action. Make sure the cover shot operator has a focused picture at all times because you might be going to him at any time.

If you use the SEG a lot, it's good practice to think out "problem" situations and ways to cope with them. Not long ago, I encountered just such a "problem" on network TV. I was watching a baseball game when, for some unexplained "technical" difficulty, all but one camera went out of commission. Fortunately, the one camera that was working was situated behind the plate. The next couple of hours consisted of two basic shots. They were mostly long shots with an occasional very cautious zoom to a medium shot. Fortunately for the network, the commentator, an experienced radio man, was able to carry the show with some clever dialogue.

DIRECTING THE SHOW

Directions to camera operators must be clear, brief, and to the point. Sometimes the director might want to let his crew know more specific details about setting up specific shots, but a good director will always thoroughly brief his crew *before* the production. It's quite simple. The better the dialogue between director and camera operators, the less the director must worry about camera movement and the more time he has to think about what the camera is seeing. In most cases, the conversation via the interphone is one way. Nevertheless, there are times when camera operator "talk-back" is important. A one-way intercom, though less expensive, does have some drawbacks.

It will facilitate the director's ability to communicate quickly and effectively between cameras if he assigns a number to each camera. A camera number should be clearly displayed at several points on the camera so that both the operators and the director can see it. The director should also attach the camera numbers to the appropriate camera preview monitors.

The variety and extent of camera directions will depend on the experience of the camera operators. Time spent in rehearsals and training will greatly alleviate the frustrations and pressures one so often encounters when using inexperienced camera operators. Once again, the key is preproduction planning. This includes meeting with the crew beforehand, agreeing on terminology, and demonstrating what you mean when you give a specific direction.

When I train camera operators, I first teach them the follow-through focus. Then they are assigned a specific area and no more than two general types of shots—all within a follow-through focus range. Sometimes it's hard to remember what it is like at first to be put behind a camera and told to do "just as I say." I recall having a great fascination for what I was doing, but very little understanding as to *how* I was doing it! So take it slow with new camera operators; get them working for you—not against you.

SCRIPT DIRECTIONS

The use of verbal terms between director and camera operator, such as "close-up", "tight shot", "pan right," is an essential part of coordinating a multicamera production. These terms are just as essential in scriptwriting. They make the director's job much easier. It's not easy to script camera directions. You must first understand what the camera can do and what the SEG can do. Even so, things never look the same on camera via the SEG as they looked on the script. When developing a script with SEG directions, allow a day or so for dry runs. At that time you can make modifications or adjustments. The following is a list of some of the common abbreviations for specific camera directions. Keep in mind that audio as well as video directions will be needed in the finished script.

XF—cross fade, actually a dissolve from one camera to another.

CU—close-up (be sure your camera person knows what you call a close-up).

FTA—follow the action (used a lot when scripting the coverage of uncontrollable events).

FU—fade up from black.

Cut to 1—switch to camera one.

TL—truck left.

You will have to make up your own list of directions for both written and verbal commands. How experienced is your camera crew? Consistency is the key. Make a list of directions and basic procedures and stick to them.

The most effective way to get script directions from the director to the individual camera operators is with an intercom system (preferably a two-way system). With good directions from the director, a camera operator will seldom need to "talk back," but there are times, particularly with inexperienced camera operators, that questions must be asked. Most SEGs now on the market have built-in intercom systems or provisions for connecting an external system. It should be an important consideration when you are shopping.

If you already have an SEG with no intercom, you have a number of alternatives. You can buy a completely separate system.

Such a system usually requires another cable to the three umbilicals that connect the camera and the SEG. Some of these systems can be integrated into your existing system using an open or available cable in the camera cable. Your VCR dealer can assist you with this type of setup. There are a number of relatively inexpensive wireless intercoms now on the market. Be sure to check them out. If you can't afford an intercom system, another alternative is to set up a program monitor or a second monitor that displays what's going on "on-line." This type of monitor (you can use a regular TV set coming rf out of your VTR/VCR) will allow the camera operators to see when their camera is "on-line." For this system to work, the director should give his crew advanced directions as to who covers what area. This might mean assigning one camera to follow the action with medium to close-up shots while a second camera is told to cover the entire scene with long or general shots. Once again consistency is the key. If you have a plan and get a way from that plan, watch out!

PUTTING IT TOGETHER

A primary advantage of helical-scan equipment is its mobility. Adding cameras and monitors—all the gear needed to put together a multicamera system—can quickly create a cluttered environment and a system that is clumsy and difficult to use. There are two ways to avoid the cluttered syndrome (sometimes referred to as "spaghetti city") and both involve consolidating the gear. You can buy a ready-made production console. Just make sure it offers maximum mobility. This is particularly important if you do a lot of location work.

The other, less expensive, approach is to make your console from plywood. The advantage to making your own—aside from saving $500 to $800—is that you can tailor it to your own needs. One additional piece of advice on buying a ready built console is that you should make sure that what you buy is what you need, and that all you buy is what you get. One client I know unboxed his brand new unit only to find that some of the equipment was not there and other components were not hooked up properly!

PRODUCTION TIP

If you have an SEG with a preview output, you can feed this output into a second VCR and record all of your preview shots. This way your second camera can pick up crowd shots and other shots that can later be inserted into your master tape.

Chapter 7

Incorporating Other Media

This chapter deals with techniques for incorporating photographs, slides, film, charts, and titles into your video productions. It also describes the use of character generators and computers to create titles and graphics.

Before considering the incorporation of other media, you must first evaluate your system's capabilities. Let's assume you have a basic system consisting of a camera with auto mechanical iris, zoom, a tripod, a nonediting VCR, a microphone, a TV set, and three lights.

SLIDE-TO-TAPE TRANSFER

The setup for a slide-to-tape production is simple (Fig. 7-1). Don't use your movie screen because the screen texture will be picked up by the camera. A piece of flat white poster board will do fine. There are several key factors to consider in making a slide-to-tape transfer:

TV Format. Select slides that conform to the TV format (slides that have been taken in a horizontal plane). Vertical slides can be used, but they will probably require "cropping" by zooming in with the camera. This will cut off the picture at the top and bottom. If you know ahead of time that the slides will be transferred to tape, you should take them in a horizontal plane and shoot "wide" or "loose" as well. The TV aspect ratio is 3 units high by 4 units wide. The 35mm slide aspect ratio is slightly wider. By shooting

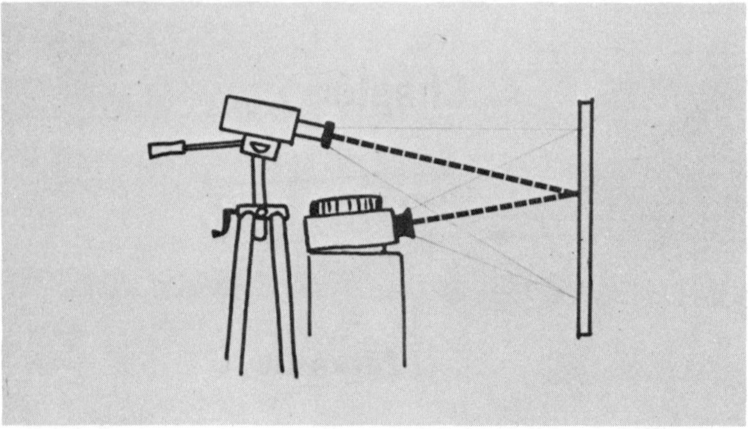

Fig. 7-1. This is the setup to use when transferring slides to videotape.

wide, you can compensate for this difference and always leave yourself a margin area or head space. This makes it easier to fill the screen with the slide and avoid showing the slide's top or bottom edge.

Contrast Ratio. In selecting slides, keep in mind the TV's limitations in dealing with contrast level differences. The less contrast in a slide the better the transfer results. Remember this if you are taking slides for subsequent tape transfer, and try to compose shots without a lot of contrast. Pay particular attention to the background. Select a background with an even contrast level. Light and color should not distract from the main subject.

Color Correction. The reproduction of the slide's color is dependent on two factors: the type of slide film and the type of projection lamp. If you have trouble recreating colors when you project your slides, try changing the camera's filter. Each time you change the filter, do a white balance on something white in one of the slides you are using. You could also check the color temperature of the bulb in your slide projector. The 3200-K bulb (used in most projectors) is perfect for the TV camera. There are several excellent films for making slides. Kodak Ektachrome films are the reversal type and are well suited to the task. A film I use frequently is high speed Ektachrome (tungsten). Outside you need a Kodak No. 85-B filter. Inside you can use tungsten TV lights with no filter.

Framing. Before taping your slide show, run through the slides and see how they are framed by your camera. Because the slide format is slightly wider than the TV format, you will have to zoom a bit to keep the bottom and top edges of the slide out of the picture. If you use the viewfinder in your camera as a reference monitor, make sure it is showing you the same area that you will see when the tape is played back and viewed on your TV set.

Transition. You're now ready to transfer your slides. Turn off all lights and project your slides. Audio can be put on as the transfer is made or it can be dubbed on later. Your camera should be in the automatic iris position that enables it to quickly adjust to the varying intensities of the light. Most home cameras, being of the single tube type, will retain a ghost image of a slide when it is flashed off the screen. Actually this "feature" helps in creating "dissolve" type transitions between slides. I normally keep the size of the projected slide small to get as bright an image as possible.

You should experiment with some simple types of transition such as defocusing right before the slide changes and then refocusing right after. Another trick to add variety to your show is to place the slide projector on a lazy Susan and "slide" the picture out of the frame. Zooming in and out and panning and tilting on the slide can also produce some interesting results. You must plan your slide show carefully and practice the camera movements.

LECTURE AND SLIDE PRESENTATION

A setup for a lecture and slide presentation is shown in Fig. 7-2. Notice that there are two lights and the lecturer and the projected slide have been isolated. The slide is projected on a piece of cardboard. The camera is set up 3 to 4 feet behind and to the side of the lecturer so that when the lens are zoomed all the way out there is a medium or waist shot of him with the slide in the upper left-hand corner.

Next, set up a follow-through focus on the slide. The depth of field created by the concentrated areas of light will give a focused wide shot of the lecturer and the slide. Close-ups of the slide can be made with zooms. Close-ups of the lecturer require resetting the follow-through focus. Thus the presentation should be designed to "feature" close-ups of the slides and wide shots of the lecturer and slides. Place a mirror next to the camera so the lecturer can see the slide without turning his head. A simple fade up with the iris can be used for the opening and a fade down can be used for the close.

DISSOLVER

A very professional and relatively inexpensive program can be produced using your camera, two slide projectors, and a dissolve unit. Use a slide projector such as the Kodak Ektagraphic Carousel

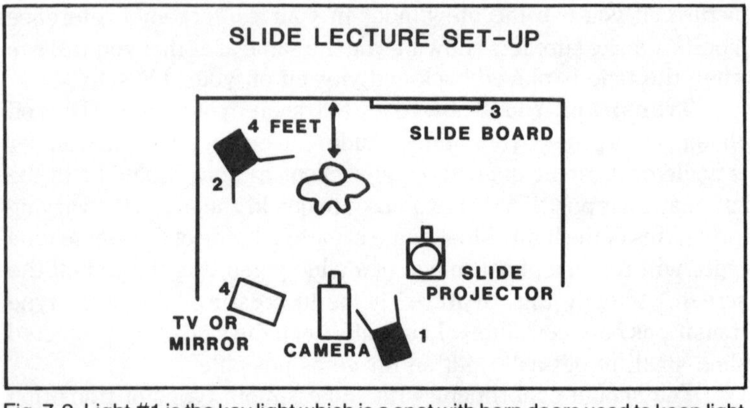

Fig. 7-2. Light #1 is the key light which is a spot with barn doors used to keep light off the board on which the slide is projected. (#3) Light #2 is a back light which is barn doored to keep light off the slide board. (#3). A mirror or monitor (#4) enables the lecturer to see what slide is being projected.

type that can be hooked up to one of the many dissolve units produced by Kodak. These units can be rented from AV houses on a daily basis for a relatively small cost. If you've never used a dissolve system, here is an outline of its features:

□ Enables you to control the forward and reverse direction of two slide projectors.

□ Creates a dissolve effect between each slide with no drop in image brightness.

□ Gives you a continuously variable dissolve rate from (½ second to up to 10 seconds)

□ Has a continuously variable automatic timer that changes the slides.

If you're going to rent a system, I suggest you go to an AV rental outfit and see what they have. Then plan your show using a light board. That way you'll minimize the rental time.

Many dissolve units can be controlled with a synchronizer device. If you create your sound track using a stereo recorder, you can then program the dissolver by recording the sound pulses from the synchronizer on the second track. To transfer this type of show to video, you must use a VCR that has two audio tracks.

HOLE IN THE SCREEN

The hole in the screen multimedia or "mixed" media presentation is something I have developed specifically for video users with a basic, single-camera VCR system. See Figs. 7-3 and 7-4. Here's

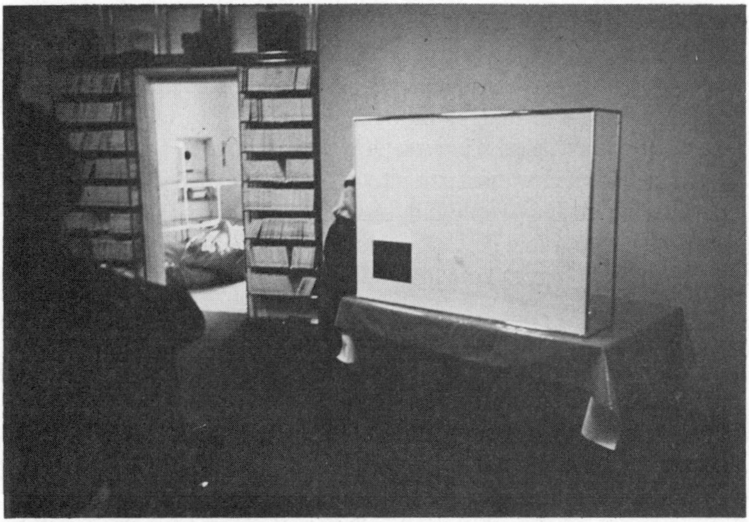

Fig. 7-3. A hole-in-the-screen setup.

how it works. A videotape is played back on a TV set. The TV is set up behind a screen with a hole cut in it that allows you to see only the screen portion of the TV. As the tape is played back, slides are projected onto the screen. The TV image becomes part of the slide image and appears to be inserted at the bottom, left-hand corner of the slide.

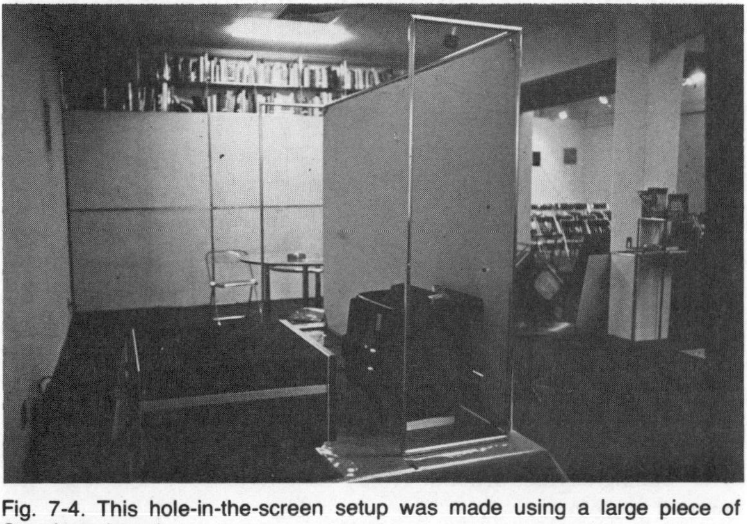

Fig. 7-4. This hole-in-the-screen setup was made using a large piece of Styrofoam board.

Make sure the hole for the TV is set in far enough from the edges of the screen so that it will "fit" nicely into the slide being projected on the screen. I have used two types of material for the screen: ¼-inch plywood, painted flat white, and Styrofoam art board that can be purchased from graphic supply houses.

This adds a new dimension to a "talking heads" videotape or to a program featuring people talking about objects or places that have or can be photographed. The videotape can be made with a single-camera system. An audio/video fade between slides can be used as a transitional device.

Keep your interviews short for this type of presentation (not more than five minutes). When you take the slides to support the subject being presented, remember to frame your shots with the "hole" in mind. When using slides you already have, pick those that will "fit" around the hole.

If you have a VCR with two audio tracks, you can synchronize the changing of the slides with the videotape by using a slide synchronizer. This unit can be used to generate the changing tone that is placed over the unused track of your VCR. When the tape is played back, the line output is fed back into the slide changer that is hooked up to the slide projector. With this setup, your multimedia presentation is totally automated.

FILM-TO-TAPE TRANSFER

The basic problem encountered with transferring film to videotape is a flutter that is created because of the frame rate difference between the projector and the TV camera. The TV camera scans at 30 frames per second; a 16mm camera projects at 24 frames per second; and a 8mm camera projects at 18 frames per second. Short sections of film can be transferred with a minimal amount of flutter using the following method:

☐ Set up your camera and projector the same way you set up for a slide transfer.

☐ Make the projected image as small as possible.

☐ Start the projector in a darkened room. You'll have to crop a portion of the projected image because it's not in TV format. Use the automatic iris setting. This will overload the auto light sensing device and average out the picture's flicker. You can't do this with the more expensive three-tube cameras because their automatic light-sensing devices are much more sensitive. Let's hear it for the low-end cameras!

If you plan to have entire movies transferred to videotape consider having it done at a duplication facility. They use a device called a *film chain unit* that consists of a TV camera and a film projector that has been modified to produce a shutter rate very close

Fig. 7-5. A film chain can be used to tape slides as well as transfer film to videotape.

to the TV scanning rate. This unit eliminates all the flutter that normally occurs during a "straight" transfer. See Fig. 7-5.

TITLES

There are many ways to create titles for your video productions. One of the easiest is to tape ready-made titles that you find on location such as signs with the names of parks, buildings, recreational fields, or companies.

Title cards can be made using poster board and a felt-tip pen. Make your lines thick. These poster boards can then be set up and shot in the taping environment. I have also used the old "vaudeville walk-on" approach where I have a person walk on and off the set holding an oversized title card.

You can affix titles to glass using the vinyl letters available at graphic supply houses. Hold the glass close in front of the camera and start your tape. Focus first on the titles and then on your taping environment, panning or tilting into the action.

If you have more than one title, make them up on 3 × 5 file cards using press type or a felt-tip pen. Then place them on a lazy Susan or a turntable. Mount the camera over the turntable. Fade up, revolve the turntable, and, at the end of the titles, fade down.

A "vertical crawl" can be created by placing titles on a long vertical piece of poster board or other similar "stiff" backing. The camera is then mounted on its side in front of an easel. The vertical card is placed on its side on the easel and "pulled" in front of the camera. Leave yourself plenty of leader on your title card, both in front and at the end, for fading up and down. I find this method of pulling graphic material easier to use than panning or tilting the camera.

If you have a computer, such as the Commodore VIC 20, that hooks up to your TV, you can feed the computer's rf directly into your VCR. Now you have a character generator you can use to create a wide variety of titles and graphics.

ARTWORK

Artwork is a general term that describes charts, graphs, drawings, and water colors. Many of the methods described in the section on titles can also be used to show artwork (turntable, pull across, vaudeville walk-on). In taping artwork, it is very important to standardize dimensions. This will enable you to tape at one camera distance and focus. An item I have found very helpful in preparing standardized artwork is a TV graphics production template made by Kodak that can be purchased at most camera stores. See Figs. 7-6 through 7-9.

If you are an artist and want to incorporate your own drawings, be sure to draw them with fairly thick lines. Felt-tip pens produce excellent colors for TV. Water colors are good when used as a background wash.

Keep charts and graphs simple. Use large lettering and solid colors. Remember to prepare your artwork using the TV format described at the beginning of this chapter.

EXPANDING YOUR SYSTEM

There are several items that can be added to your basic system to enhance your creative abilities when using the new techniques you have just learned.

Tripod. Chances are your present tripod isn't that good.

"Wait a minute!" you say. "I paid $40 for this tripod!"

If you're going to do a lot of graphics, you should expect to spend at least twice that much for a good tripod. The key element is the head (Fig. 7-10). You need one that allows you to make smooth pans and tilts. Consider buying a fluid head. Although expensive, it has some excellent features. This type of tripod allows you to level artwork at the camera head rather than trying to pin or tape it on one

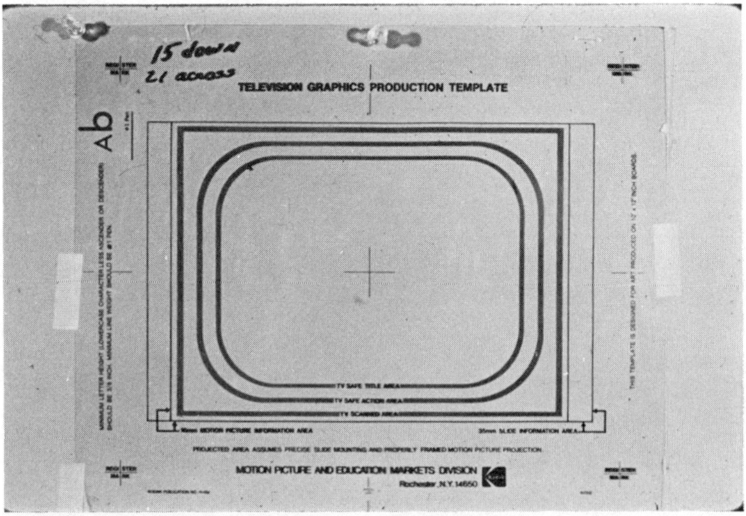

Fig. 7-6. Kodak sells a transparent template that is very useful in preparing graphic titles and other artwork for TV. Ask your local photo dealer to order one for you.

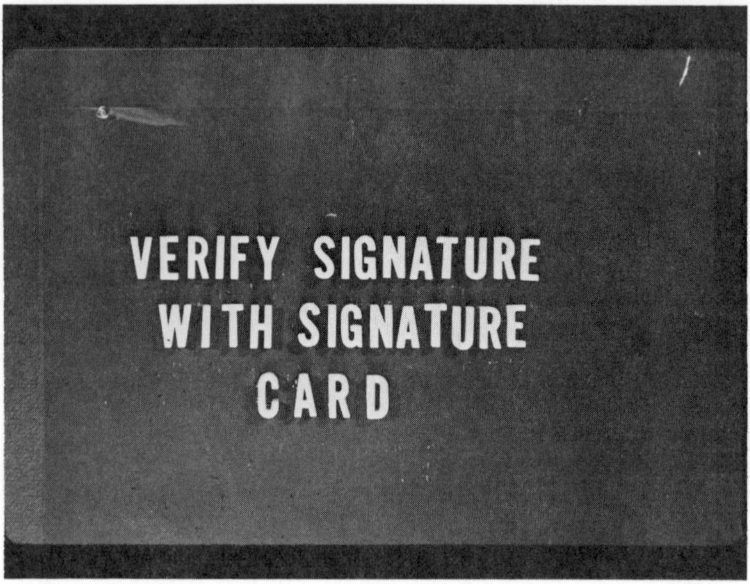

Fig. 7-7. Magnetic letters can be used to create titles with a dimensional effect.

side or the other. When selecting a tripod, be sure to take your camera along so you can try it out on different models.

Close-Up Lens. Many new cameras have a close-up or microfocus position on the zoom lens. This allows you to focus on

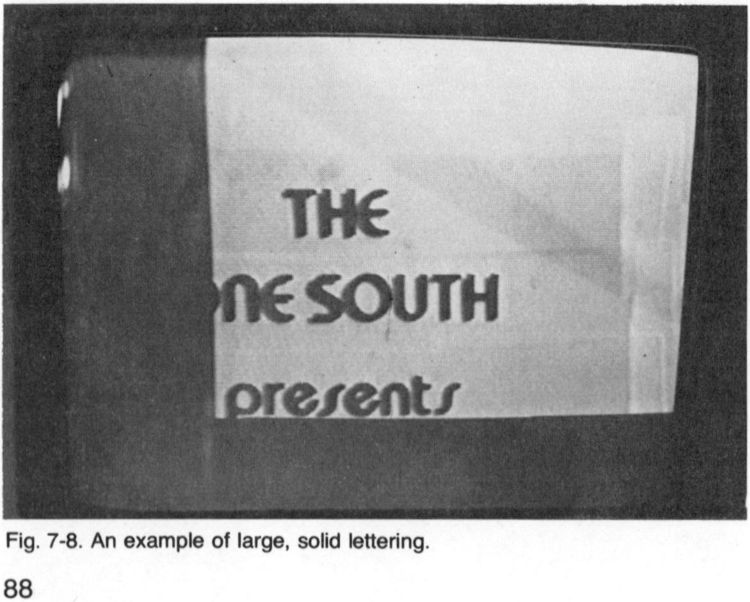

Fig. 7-8. An example of large, solid lettering.

Fig. 7-9. This title was created with an SEG, and keying graphics were created using press-type.

objects just inches away and opens up a whole new dimension in graphics and in general production coverage and subject selection. Being able to get that close to an object means you can introduce everything from postage stamps to ants on your tapes! The close-up

Fig. 7-10. A good tripod head is an important factor in shooting graphics. This head enables you to pan, tilt, and slant the camera.

Fig. 7-11. This professional character generator that features a built-in disk drive that enables the user to store many pages of graphics. Approximate cost of a system like this is $15,000. That is perhaps a little too high for the average home video user! A home computer can also be equipped with a disk drive for storage and used in the same way for much less money (under $1000).

lens can either be mounted directly onto your camera or onto a lens adjuster that fits on the end of your camera lens.

Lens Filters. Several manufacturers produce filter sets that can add a variety of effects to both your graphics and your productions. The filter kit comes with an adapter that fits right on your camera lens. With these filters, you can create color diffraction, star burst effects, multiple images, and the softening and coloring of your entire picture.

Genlock Device. Genlocking is the ability to lock up the sync from a prerecorded tape or live signal with the sync of a black-and-white camera signal. Basically what happens is that the composite video signal from the prerecorded tape or live source is fed into the genlock. There it's sync is stripped and used to drive the black-and-white camera. A genlock device will enable you to put titles and special effects into prerecorded tapes. Sony and Panasonic manufacture units with genlocking capability for the small video user at an affordable price ($400 to $500 including a black-and-white camera). These units have the following features.

☐ **Fades.** Fades can be produced in prerecorded tapes. This allows you to fade out or up into a scene.

☐ **Colored Backgrounds.** You can generate colored backgrounds. You can key (one video signal is controlled by the waveform of a second video signal when they are combined to form a composite picture) letter or symbols picked up by the black-and-white camera into the colored backgrounds.

☐ **Insert Titles.** Titles can be keyed into prerecorded tapes giving a very professional look. This enables you to create opening titles and to insert subtitles into prerecorded footage.

☐ **Reversals.** Images can be reversed and colored. Try feeding a colored camera signal into the unit. By focusing the black-and-white camera on the same object the color camera is focused on, you can colorize the images and produce some very interesting results.

A genlock unit will greatly increase your production capability. This is particularly true in post-production editing. Probably the best way to create titles is by using press type. Press type works particularly well when keyed over artwork using a genlock device. When using press type to do video graphics, keep in mind the points outlined in Fig. 7-14. Beware of letters that can cause problems. Some of the examples shown in Fig. 7-14 might work fine for you, but be sure to check them out. Get to know what types of typefaces will work with your cameras and matt unit. Also beware of logos that will cause problems.

Video Feedback. Video feedback was one of the first techniques used to create "video art." You hook up your system and then simply point the camera at the TV screen. You might have to zoom in and out and tilt the camera from side to side. Eventually you'll create "colored waves" and "spiraled" effects. Using a color camera to create video feedback and then feeding the signal into a genlock device, you can create all sorts of interesting backgrounds for titles.

Character Generator. A character generator (Fig. 7-11) is a device that allows you to generate electronic characters in a variety of fonts and colors. These characters can be keyed into a live camera system or, by using a genlock device, keyed into a prerecorded tape. Currently there are only a few "affordable" character generators on the market for the home or small system user, and their capabilities are somewhat limited.

Some cameras now have "built-in" character generators that let you add titles to your pictures. Though convenient, I find these

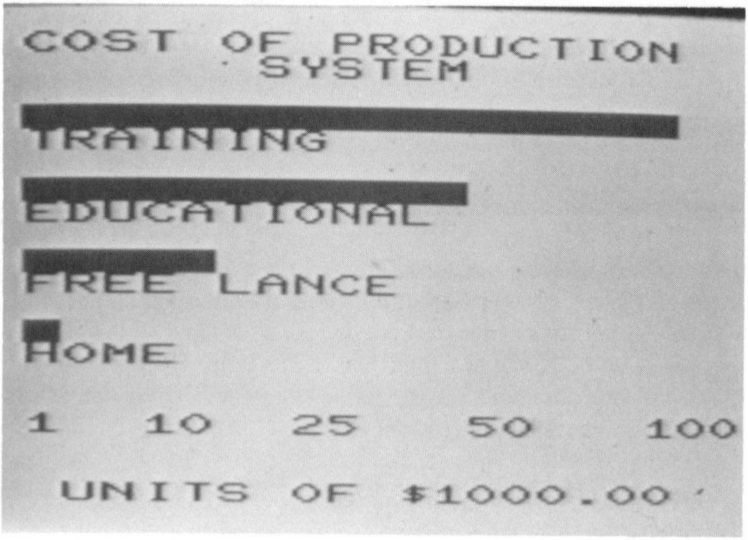

Fig. 7-12. A home computer was used to create this graphic.

quite limited and slow to work with. The more sophisticated units have features such as different size characters and fonts, background generators, spacing and centering controls, crawls and rolls, and erasing and editing, as well as page access and storage.

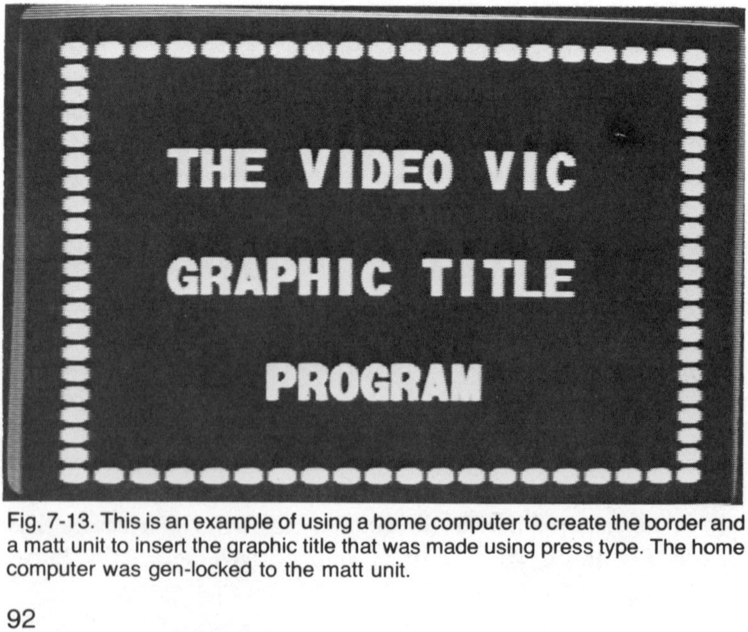

Fig. 7-13. This is an example of using a home computer to create the border and a matt unit to insert the graphic title that was made using press type. The home computer was gen-locked to the matt unit.

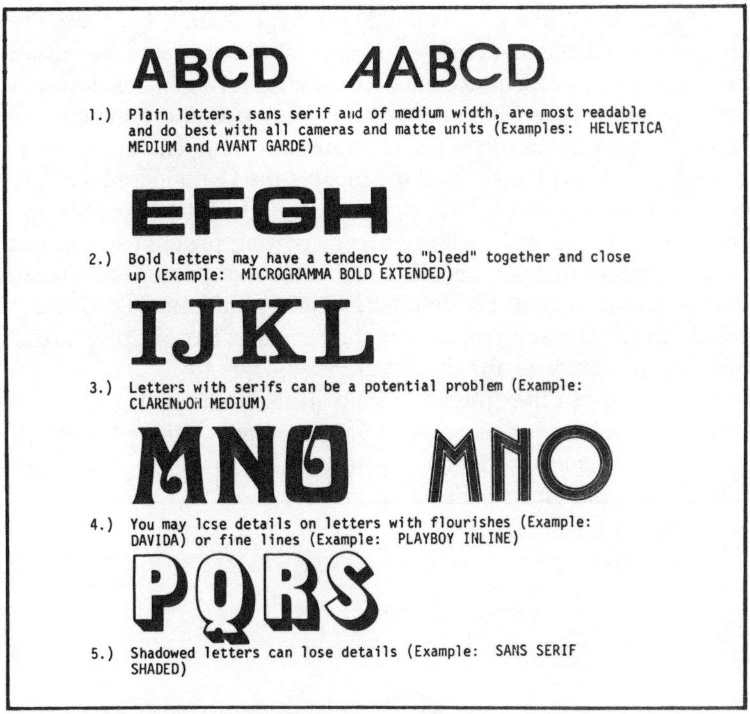

Fig. 7-14. Video graphics guidelines.

These units are very expensive. I find that the home computer can provide many of the same features at far less cost.

Home Computer. Many home computers can be used to create titles, charts, graphs, and other special effects for your video production (Figs. 7-12 and 7-13). The key is to select a computer with a signal compatible with your VCR. If your computer connects to your TV set, then chances are you can feed it into your VCR. Just hook up the computer's rf output into your VCR's rf input and tune your VCR to the channel output of the computer (either 3 or 4). Any images or sound you create with the computer can be recorded on your VCR. Many manufacturers will provide adapter cables that will enable you to send out straight audio and video line signals. This will greatly increase the quality of your video and audio signals and provide you with sharper images and cleaner sound. If the actual video signal strength is less than 1 volt, you might want to buy an amplifier that will bring the video and sync up to the proper level. This is particularly important if your tapes will be played back on cable or be broadcast.

Most computers give you only one type of font, but there are programs available that enable you to create your own characters. I have found that by feeding computer signals into a genlock device, I can use press-type characters. I can also use the computer to generate a wide assortment of stationary and animated backgrounds. The unit I have used the most is the Commodore VIC 20. Graphs, charts, and titles are created simply by hitting keys on the keyboard. With a little experience in writing programs, you can create graphics that you can store, animate, and update. The VIC 20 also generates sound. I have created background music and sound effects with it. I am convinced that the computer has exciting and far reaching possibilities for the small video user.

The use of other media in your productions will greatly increase the impact and meaning of your message. But be careful! If not properly planned and implemented, your message could confuse and distract the viewer. See Fig. 7-14.

Chapter 8

Video Grammar

Portions of this chapter appeared in the Vol. II, No. 1, January, 1979 issue of *Educational and Industrial* magazine, published by C.S. TEPFER Publishing Co. Inc., 51 Sugar Hollow Road, Danbury, CT 06810.

In the preceding chapter, I discussed the technical requirements and approaches to editing videotape. This chapter covers specific techniques you can use to give your edited productions continuity, rhythm, and interest.

Any video user with a limited budget and limited equipment can improve his productions by watching how it's done by the high-budget, equipment-rich shows on the commercial networks. But to understand what you're seeing, you must be visually literate. You must be able to understand and recognize the basic grammar of video. With this knowledge, you can watch commercial programming with an educated eye and, I think, you will find that you will enjoy it much more.

For the home video user, some of the questions are as follows. What shots did they use and in what order? Why? How do they get from here to there, what transitions did they use and how effectively? What can I adapt for use in my own productions? How?

To help you understand what you see and use what you learn, this chapter describes the elements of visual grammar. They are camera movements, angles, and transactions that you can combine to give your programs pace, rhythm, and impact.

CAMERA MOVEMENTS

Follow-Through-Focus. Before you learn how to move the camera itself, you must learn what to do with the zoom lens. The most basic technique is to be able to zoom in and out while keeping your subject clearly in focus. Like all camera movements, it requires practice and more practice until it becomes second nature to you.

☐ Step 1. Zoom in as closely as possible on the subject.

☐ Step 2. Focus your picture by using the manual focus ring on the lens barrel.

☐ Step 3. Zoom out. You should have follow-through focus now. Please note that this does not mean that you can now zoom in and out on any subject and maintain focus. It does mean that you will maintain focus on objects that are the same distance from the camera as the subject used to set up your initial follow-through focus.

Dolly. Moving the camera closer to or further from the subject—with a wheeled platform (dolly) or, if you are using a hand-held camera, by moving yourself (you are the platform)—is an acquired skill. The effect might seem the same to you as using a zoom lens, but actually it's quite different when viewed on a TV screen.

If your camera is mounted on a dolly, this maneuver requires two people (the camera operator and another person to guide and steady the dolly). Prior to executing a dolly move, make sure the surface over which you will move the dolly is free of debris such as pebbles, cables, and bottle caps. A broom is standard equipment. Make sure that the wheels are all pointing in the same direction so that, when the dolly is moved, your picture doesn't suddenly jerk to the left or right. If *you* are the camera platform (when using a hand-held camera), brace your arms tightly against yourself to provide the necessary stability. Otherwise, you'll get jiggle, bob, and weave in your picture.

Pan. Swinging the camera horizontally without moving it out of line is called a pan (Fig. 8-1). This is used to follow the movement of your subject as he walks across the set, room, street, or whatever. In general, a pan should go much more slowly than you think. So slow down. Remember all the too-rapid pans you have seen and the "home movie" effect that resulted.

Steadiness is the watchword here and practice is the key. Give your subject "lead" or "nose" room by keeping slightly ahead of the action. A moving subject should not be in the center of the picture. There should be more space ahead of him (i.e., in the direction he is traveling) than behind him. Positioning a moving object in the center of the frame makes it look crowded for space. This gives the impression that you can't keep up, and it will get away (escape out of the picture). There might be times when this kind of effect is desirable.

Tilt. Moving a stationary camera up and down vertically is called a tilt. Executing a smooth tilt requires the same steadiness, slow pace, and lead space as a good pan. Take more time than you think you need. And practice!

Truck. A truck moves the camera horizontally as the subject moves. This gives the effect of walking down the street. Although the motion might seem the same as a pan, the effect is different. Remember the "nose" room!

Trucking can be done with the camera mounted on a dolly, and with the second person plying that broom to sweep away twigs, beer cans, and anything else that might disturb the smooth progress of the wheels. When handholding the camera, brace your arms and "walk with" the camera. Practice so that your walk is a smooth

Fig. 8-1. Hand-held camera shots can be very shakey. You should hold the camera firmly against yourself. Make your pans and tilts by moving your entire body in a slow and steady movement. Practice your moves several times before rolling the tape.

Fig. 8-2. Camera point-of-view is a key to developing a picture that focuses the attention of the viewer. Getting close to the action is important in establishing a good camera point of view.

glide. A second person is useful to guide you or your dolly. Don't bounce up and down when your heels hit the ground. Slow and steady does it.

Arc. This is very useful to establish spatial relationships. Take a pretty girl looking over a porch railing. Do a 360-degree arc, and the audience discovers she is looking out over a cliff to the ocean, with nothing but deserted landscape to either side of the house. Do a 180-degree arc and discover somebody scrambling over the edge of the cliff. Suspense! Drama! The arc has established the solitary location of the house and the apparent vulnerability of the girl. In more prosaic contexts, an arc can establish that the action takes place on a busy street corner, or that you are standing in the middle of a vast, empty athletic stadium.

Depending on the pace of your production, an arc can be done somewhat more quickly than a pan. Nevertheless, don't speed it up too much or the audience will lose detail and the scene will blur. Fast pans or arcs can be used for transitions. Keep it steady and don't let the camera tilt unless you want a roller-coaster effect, and the resultant vertigo induced in the audience.

POINT OF VIEW

Camera work can establish the focus of the viewer's attention or it can provide psychological tensions that will add impact (Fig. 8-2). Done poorly, however, it can also confuse. In general, the closer the shot, the more immobile the subject and the camera must be. Even tiny movements of either will be greatly exaggerated in all close-ups. The closer the camera is to the subject the larger the movement will seem. So you must have either a rock-steady subject or an extremely alert camera operator.

Eye Level. Eye level is the most common camera angle. The eyes of the subject should be about two-thirds of the way up the screen. If you are doing something that is mostly shots of people, put a piece of tape across the monitor screen two-thirds of the way up from the bottom, and take the shots of your subjects so that their eyes remain at that level regardless of how far away or close to the subjects the camera is or seems to be.

If you must "crop" a shot, crop from the top. Better to keep the subject's eyes at the two-thirds level, and lose part of the top of his head, than to move the camera up so you catch the crown of a hat but drop the eyes to the middle of the screen.

Low Angle. With a low-angle shot, the camera is well below the subject's eye level and shoots up. This angle can add drama. It makes the subject loom at the viewer and often gives the impression that momentous events are about to take place. Used subtly, it can add importance to the subject and make him appear slightly larger than life. Look at political commercials, for example, or speeches from an elevation. Used not so subtly, the low angle can have a negative effect—making the subject seem domineering, threatening, and massive.

If you are using a portable camera, the low angle might call for some acrobatics. Rather than hunkering down on your haunches, lie flat on your stomach. This permits you to brace your elbows on the ground and provide a steady base for the camera. Or kneel on both knees and brace against them. One advantage of video is that you don't have to always get down to the eyepiece of the camera. You can study or set your camera angle by monitoring the shot using another monitor hooked up to the camera or to the VCR's output. This allows more flexibility not only in low level, but in high and extreme overhead shots as well.

High Angle. With the high-angle shot, the camera is above the subject's eye level and shoots down. This gives a squirrel's-eye view and minimizes the importance of the subject. It makes him look smaller and often weaker as well. A parent scolding a child, or a sentence pronounced from the raised bench of a judge are examples.

The high-level shot can also be used to show the subject in relation to his surroundings or to focus the attention on what he is doing with his hands or feet.

You might have to use some agility and imagination to find a perch from which to shoot. The top of a car works well or you could borrow a ladder. Be sure you are secure enough so that you don't have to use your hands for holding on to something. You need your hands to brace the camera.

Extreme Overhead. With the extreme overhead shot, the camera is directly above the subject and shoots down on it vertically. For informational programs, this is useful to show something on a flat surface (a map, for instance). Many classrooms used as studios for educational programs have a camera permanently mounted to shoot down at the lecturer's desk so that he can write formulas, display graphs, or simply doodle to provide graphic illustrations for his talks.

If you can get enough light on the subject, a mirror shot can be used. Sometimes a mirror is easier to position than the camera. Keep in mind that the mirror will *reverse* the image. To shoot an extreme overhead shot of a street crowd celebrating Mardi Gras, borrow somebody's second-story window.

Any of the preceding points-of-view can be combined with any of the following types of shots. There is a wide selection of possible ways to show your subject.

Wide Shot. The wide shot takes in the entire subject and its surroundings. This is an orientation shot that tells the viewers the location of the scene. On the street, in a field, in somebody's office, are examples. The wide shot is often called an "establishing" shot. You also need it to cover broad action involving several people and a fairly large area.

Medium Shot. The medium shot is the utility shot in your arsenal. It focuses attention on one or two persons or things at conversational distance (i.e., shows a person's head, shoulders, and upper torso). The environment is secondary in a medium shot. Presumably the audience knows where they are and is ready to settle down to more important matters. The medium shot is also useful as a signal that a sequence that has involved a lot of close-ups is now over.

Close-Ups. The close-up shot is for details. The content of the note your subject found propped on the mantel, a facial expression, what "folding" is in relation to cake batter, or that formula on the lecturer's desk are examples. By cutting out the unwanted portions of the image, the close-up demands that the viewer concentrate on what is left.

Extreme-Close-Up. A closer close-up, the extreme close-up shot is used for dramatic effect. Examples are, an open mouth screaming or a surfer reflected in a pair of sunglasses. The extreme close-up also provides information. Examples are a needle's movement through the stitches of petit point or the calibrations on a fever thermometer. The camera seems to be (and sometimes is) only inches away from the subject. This will most likely require lens extenders because this shot is not possible with standard lenses.

Depth of Field. Using depth of field is a way of focusing attention by sharpness of image. The camera operator shoots the object he wants the audience to see clearly in sharp focus while leaving the rest of the picture fuzzy. This implies that the other items are not as important. The clear image can be anywhere—the foreground, middle distance, or background—as needed.

Not only is this technique useful as a change of pace, it also lets you and the camera remain stationary while your subject moves toward or away from you. You can follow a deer as it bounds away from you through the trees, you can pick out one flower from an entire bed, you can select one face in a crowd, or you can follow the progress of a rat running through a maze. Or you can shift the focus of attention without any movement of the subject by showing the foreground tall grass clearly, for example, and then shifting the depth of field to show the bird on its nest behind the grass shield.

THE EYELINE

The eyeline is an invisible (or actual) line through the center of your subject and perpendicular to the head-on camera angle (Fig. 8-3). It is used whenever you are shooting one sequence from two or more different points of view, either by using one camera for each angle or by moving a single camera to different angles between shots. The purpose of the eyeline is to keep the same perspective when shifting angles. For that reason, *never* cross the eyeline with the camera(s)!

Take a simple interview with two people: Sam and Joe. As they face each other, the eyeline passes through them according to the diagram (Fig. 8-3). If it will help, actually draw a line on the pavement, or in the sand, or lay down a strip of tape on the floor. The camera angles are from the right, to pick up Joe's face, and from the left for Sam's. The viewer will see each of them in a three-quarter perspective.

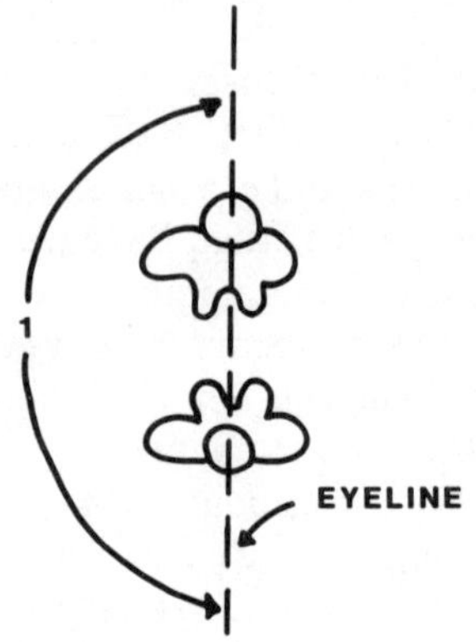

EYELINE

THE TWO PEOPLE CAN BE COVERED FROM ANY POINT IN THE ARC (1)

Fig. 8-3. When shooting two persons for subsequent editing or when using a two camera system, be sure not to cross the "eyeline."

Sam asks Joe a question. To the audience, Sam appears to look off to the left side of the screen at Joe. Joe answers and the new camera angle shows Joe looking off the right side of the screen toward Sam. It's very familiar and very comfortable as long as the camera remains on the *same* side of the eyeline.

If the camera operator crosses the eyeline to shoot Sam asking his umpteenth question, what happens? Suddenly the perspective changes and Sam is looking off the right (wrong) side of the screen. But you have already established that Joe is to Sam's *left* and here is Sam looking in the opposite direction. What happened? The audience will be disoriented and wonder how come Sam and Joe suddenly decided to change places. They didn't. The camera is in the wrong place. The same thing happens when shooting action. An example is a horse race. From the stands, you shoot the horses getting into the starting gate, the gate clanging open, and the first jockeying for position as the horses run to the viewers' right. If at that point, you cut to a second camera positioned in the grassy oval on the other side of the track (across the eyeline), suddenly the horses will be running to the viewer's left. You can take shots of the backstretch from the grass because, by then, the eyeline has shifted from in front of the stands to the far side of the track. Watch your eyeline. Pick the side you want to shoot from and stay there for the duration of the sequence.

TRANSITIONS

The most often used transition is the simple jumpcut. It is an abrupt change from one camera angle or point of view to another. But you can't cut your way through an entire program without monotony or real disruption to continuity. The following are some transitional techniques you can use to provide interest and variety.

Matching Action. Matching action is a method of breaking an action sequence so you can shift camera angles or points of view to provide variety, rhythm, and pace. See Figs. 8-4 through 8-7. The easiest way to explain it is to give an example. Suppose your talent is seated at a desk. He announces that it is now time to go outside. He gets up and walks over to the door. He opens the door. The scene now switches, or "cuts," to the exterior of the building, and we see the talent coming out of the door to the outside. Let's rewind and explain.

First, tape the talent as he gets up. Follow him as he opens the door and exits. Now, rewind and make note of what hand was used to open the door, which foot stepped out first, etc. Closely study the movement or "action" of the talent. Instant replay makes this a real piece of cake. Now reposition your camera setup to the exterior shot. Have the talent walk, from the inside out, going through the same action or movement he performed in the interior sequence.

Now the rest is up to you as to where you want to go to match the action or edit together the sequence. It could come just as a hand is placed on the door or just as footsteps outside. By matching action, you will ensure smooth transition from the inside to the outside shot.

Screen Direction. Screen direction is important when the subject moves off camera out of one shot and on camera into the next shot. In such sequences, be sure that the subject moves in the same direction both times.

For example, a lecturer wants to move from his lectern to the graphics stand on his right. You could, of course, truck or pan and follow him as he walks. For variety (and simplicity) you decide to let him walk out of the lectern shot and into the graphics stand shot.

Be sure that, if he is walking right (off the left side of the picture) when he leaves the lecturn shot, he is still walking right (in from the right side of the picture) when he enters the graphic's stand shot. Otherwise, (like the horses) he'll seem to be going the wrong

Fig. 8-4. An example of matching action.

way. Also, be sure not to cut or edit to the next sequence until the lecturer is completely out of the shot or off the screen.

Leading. Leading can be a transition or simply a continuation

Fig. 8-5. A matching-action shot from the front of the subject.

Fig. 8-6. The camera point of view was moved to the back of the subject.

of the same shot with a refocus of the viewer's attention. Again, let's take an example. Grandma is in her living room, sitting on the sofa, reading a book. Suddenly, she raises her head and looks to her right. The camera follows the direction of her glance (she "leads" the camera) and discovers—a telephone on the desk. The implication is the phone is ringing.

Walking Through the Camera. Walking through the camera is a familiar, but still effective, transition in which the subject walks toward the camera until he seems to block out the view completely with his body, and the picture goes black for lack of light. In the next shot, the camera begins with black and focuses gradually on the subject's back as he walks away into a new setting.

Defocus/Refocus. A defocus or refocus can be used for either a change of scene or a passage of time transitions. At the end of the scene, gradually defocus the lens so the image becomes completely unrecognizable. At the beginning of the next shot, refocus on a new scene. Or defocus on a scene at high noon and refocus on the same scene at sunset, and, behold, six hours or more have passed. In the studio you can do this with lights. Out of the studio, you'll simply have to wait for sunset.

Sometimes you can accomplish both a change of time and a change of place with the same transition. One useful device here is to use the face of a clock. Defocus on an electric wall clock that reads 2:20 and refocus on the face of a desk clock that reads 4:45. The audience realizes that not only have more than two hours passed, but they are also now in a different place.

Blank Surfaces. Use what is, in effect, a patterned or solid color surface between sequences. This does not require a change of camera position, but you can shift the camera if needed. For example, if you are shooting outdoors, tilt the camera up and show the sky for a few seconds. Then set up your next shoot and begin with the camera on the sky (blue sky to blue sky). The next shot will start on the sky and tilt down into your next sequence.

Swish Pan. A swish pan transition changes the point of view for a continuing action more often than it changes a scene or time. Which is not to say that it cannot be used for either of the other two as well. But for our example, let's use a sequence in which there is no passage of time or major displacement in space.

You want to show an employee on his way to complain to the boss. Shoot the subject getting up and walking towards his office

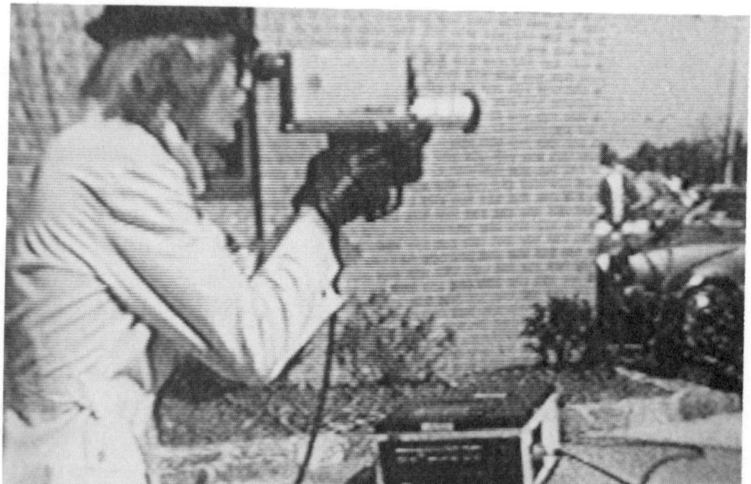

Fig. 8-7. The subject repeated the action performed during the frontal shot. Later in the editing process, the two segments were edited together at the point where the subject started up the VCR. Camera position in both the front and back shot were kept to the subject's right so as not to cross the eyeline.

door. Just as he reaches out to open the door, do a quick pan from the subject and an arc of at least 180 degrees. Then set up your next shot with the subject walking up to the boss's door. Do a quick pan, this time toward the subject, and have it stop just as he reaches his hand out to open the boss's door. The two quick pans create a swish, or blurred, transitional effect.

AUDIO GRAMMAR

Up to this point, I have focused on the "video grammar" of TV production. Audio also plays a key role in the logical manner in which a program is put together. There are a variety of audio effects that can be used to provide rhythm and continuity to your productions.

Audio Transitions. One of the easiest ways of maintaining the viewer's interest when cutting from one picture to another is to use audio transitions. This can be done with music, sound effects, or narration. The most common use of audio transitions is for the opening and closing titles.

Audio Sweetening. If you are editing a scene in which there are numerous cuts back and forth among several people, you might be faced with a sound tone problem. Part of the problem can be solved using an audio processor. You can also sweeten it with added room presence, music, or sound effects. The added sound will usually cover up the difference in audio tone.

Sound Effects. Enough can't be said about the importance of sound effects. A good sound effects tape is a must for the home video producer. You can make your own tape or buy one. Many libraries have sound effects tapes: you should check one out and listen to it.

Sound effects are usually easy to add to a production using audio dub after the video has been edited. They will add impact to your message and provide you with easy-to-use transitional devices. There is nothing like a good laugh track to liven appropriate interviews.

These are only some of the many possible techniques used to provide interest and creativity in your productions. One final suggestion is that the next time you watch a show on TV, make a list—prior to watching it—of the transitional devices outlined in this chapter. As you watch the show, keep score of the audio and video devices used. Allow room on your score card for new entries. No doubt you will come up with many more.

Chapter 9

Preproduction Guidelines

Any piece of videotape with sound, picture, and sync is a "production." To create it, someone had to point the camera and depress the "record" button. But usually we mean something more specific when we speak of a "video production." Implicit in the term is the concept of intent.

A video production is something that is made with a message and a preconceived result in mind. This should be your first consideration in the preproduction planning process. Decide what you want to say and what you hope to achieve.

There are many types of messages, but their results can be grouped into three categories: entertainment, information, or instruction. See Figs. 9-1 and 9-2.

A program that entertains affects the viewer's emotions (it makes the viewer laugh, relax, etc.). A program that informs gives the viewer facts, statements and evidence, about a specific process or event. A program that instructs trains the viewer. While watching, he learns to perform a specific skill or task.

There is a fourth category. I hope your tapes will not fall into it: programs that bore. Boring programs put people to sleep, lead them to distractions, or make them turn off the TV.

Most of the productions on broadcast TV produce a combination of results and, obviously, the most sought after result is to entertain. Most broadcast TV is entertainment oriented because audiences are so large. If broadcast TV suddenly zeroed in on

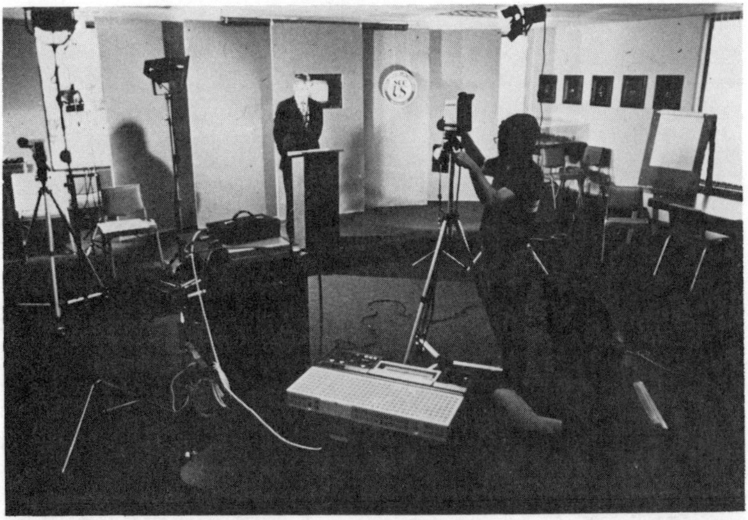

Fig. 9-1. The objective of the videotape being produced here is to inform. The presentation was taped from start to end. Note the TV screen behind the subject. It is connected to a camera setup just to the left of the set. This enabled us to show graphics on the TV screen behind the subject as he was talking. The TV screen could be brought into or taken out of the picture by zooming in and out on the subject. Several dry runs were needed before the final taping was done. This is a good production strategy for training tapes that have to be made quickly and on a low budget.

Fig. 9-2. The objective of the tape being used here was to train bank tellers. After watching a segment of the tape, the program is stopped and the trainees then duplicate the procedure they have just seen using a simulated teller station. The strategy used here, which incorporates the use of an interactive format, is very effective for skills training.

Fig. 9-3. A most important step in the preproduction process is reviewing the script with all the people involved in the production. Preproduction conferences can take place in an office or in the field (as shown here).

educational programs, they would immediately limit the number of potential viewers. Ratings would drop, and everyone knows what happens then!

Most nonbroadcast productions, from CATV to home video, do not have to appeal to the masses. They can zero in on specific needs. So be glad you don't have to make programs for the millions!

A quick analysis of your audience will give you clues as to how best to convey your message. For instance, if you are making a tape on crime prevention for your neighborhood, knowing the age level of your audience would be helpful in selecting your manner of presentation. A program featuring a police officer talking to the viewer from behind his desk might be fine for adults, but kids would probably be bored. There wouldn't be enough action or drama! Ask yourself these questions about your audience:

☐ How much does my audience already know about my subject?

☐ How big is my audience?

☐ How will my audience view the program? (In small groups, on CATV, on MATV, individually.)

☐ What is the cultural makeup of my audience?

Having analyzed your audience, you can then select the most

effective means of presenting your message. This is referred to as *program format* or *strategy*. An examination of commercial TV will reveal some strategies. Here are a few:

Strategy	*Program Example*
The interview	Johnny Carson Show
Panel discussion	Meet the Press
Documentation	Victory at Sea
Magazine	PM Magazine
News show	CBS Evening News
Drama	Charlie's Angels
Quiz show	Tic Tac Dough
Cartooning	Super Friends

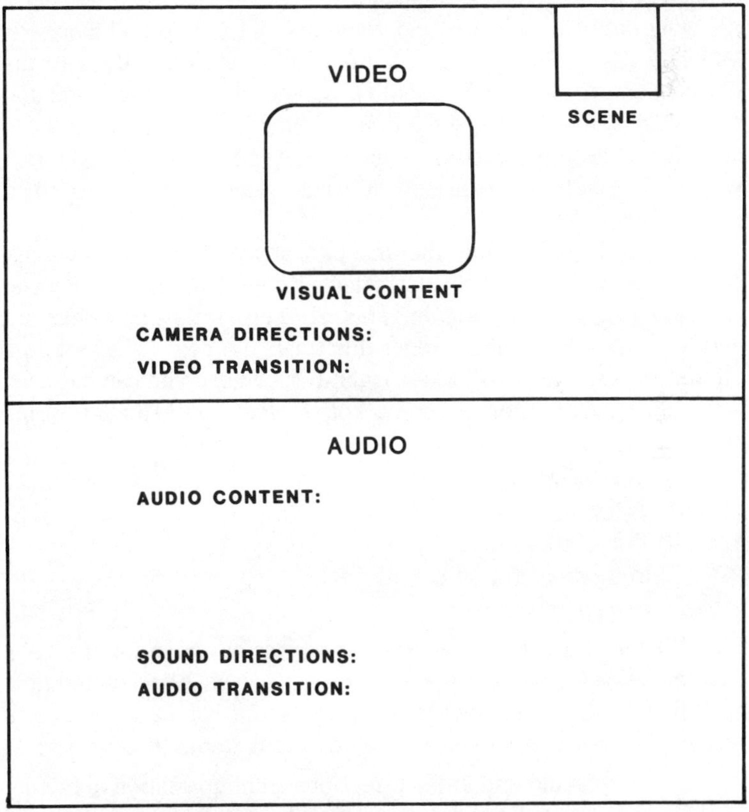

Fig. 9-4. One form that can be used is to storyboard or outline your script in a scene-by-scene sequence. Audio and video content and directions are divided, providing a clearer plan to follow when producing the program.

Once you have selected a program strategy, you should analyze it by studying the various audio/video techniques used to produce the program. This means paying special attention to these techniques:

☐ The way the show is opened and closed (music, graphics, etc.).

☐ Set design.

☐ Lighting and sound methods.

☐ Camera angles or point of view.

☐ Transitional devices (ways in which the show moves from one scene to another (use of fades, matching objects, music, etc.).

☐ Sweetening (sound effects, music, canned laughter).

Recording a program off the air and replaying it several times, is a good way to study these techniques.

One key to analyzing and copying audio/video techniques is knowing your equipment capabilities and limitations. Back to the drawing board you say? Not at all! Over the years, I have found that part of the fun of video production planning is figuring out ways to copy interesting techniques. Experiment and let your imagination run wild. You'll be surprised at what you can do even with a single-camera system.

Before launching into the final part of preproduction planning, script writing, think about your budget. Know how much money you can spend on your production. This will help in writing a relevant script. If you work out a budget this time, the next time you do a similar program you will know right away where you can cut back and where you need more money. Your budget should include items such as:

☐ Raw tape costs.

☐ Talent fees.

☐ Travel expenses.

☐ Equipment rental costs.

☐ Graphic supplies.

Now you are ready to write your script. A script is like a recipe. It is a formula for presenting your message and obtaining a specific result. There are many ways a script can be layed out or structured. Here are some of the important items to consider:

☐ Keep audio and video directions and information divided.

☐ Develop your script in a scene-by-scene sequence.

☐ Be sure to include audio and video transitional devices for

Fig. 9-5. There will be days when nothing goes right!

each scene. This will be your method for getting into and out of each scene.

☐ Include specific directions for camera moves and audio techniques. This might mean developing your own abbreviations (FU=fade up, CU=close up, etc.)

☐ Indicate approximate time for each scene.

☐ Include a diagram of equipment set up, set design notations, and props.

Now you can get ready for the production. You will need to schedule shooting times, check out equipment, and—if location shooting is called for—make a trip to the location site. Unforeseen problems will often arise during the production process (shooting the program), but you can minimize these problems by dealing ahead of time with as many details as possible. It is plain good sense to spend as much time as possible in the planning process—selecting strategies, analyzing your audience, deciding on dialogue, writing the script. It's much easier to deal with problems at this point than in the middle of an elaborate and costly studio setup or on location miles away from the comfort and security of your video den!

When your script is well planned and you are certain of your message and its desired result, you have more time and energy for the "creative" aspects of your tape. You will have more time to make sure the shot is framed just the way you want it or that the microphone is placed properly for optimum sound. Learn to control the medium so the medium will not interfere with your message. See Figs. 9-3 through 9-5.

Chapter 10

Production Models

This chapter contains a series of production activities and applications. All the activities are based on actual experiences. Let's start off with some activities that can be used at parties or for "breaking the ice" when introducing people to video. See Fig. 10-1.

VIDEO FUN AND GAMES

Broadcast Dub. For this activity you must first record one of our favorite commercials. Then, using audio dub, replace the original sound track with one of your own. Music and sound effects can be "mixed" with your narration by using a mixer or by playing music over your stereo. You can control the level of the music by raising or lowering the volume control on the stereo. You don't need to limit this activity to commercials. Try some of the popular soap operas. You'll find lip syncing short segments is fairly easy with practice. Sporting events are also fun to match with music.

Video Expressions. Creating video expressions requires a camera, a microphone, a VCR, and a TV. Prepare a number of brief directions and write them on file cards. Some examples of video expressions I've used are "make a monster face," "you just smelled something awful," "make yourself laugh," "tell us your favorite joke," and "sing a song about someone at this party." Set up your recording system in an area away from the party. Then invite your guests to come to the recording area, one at a time, and hand them a card with a "video expression" on it. Give them a minute or

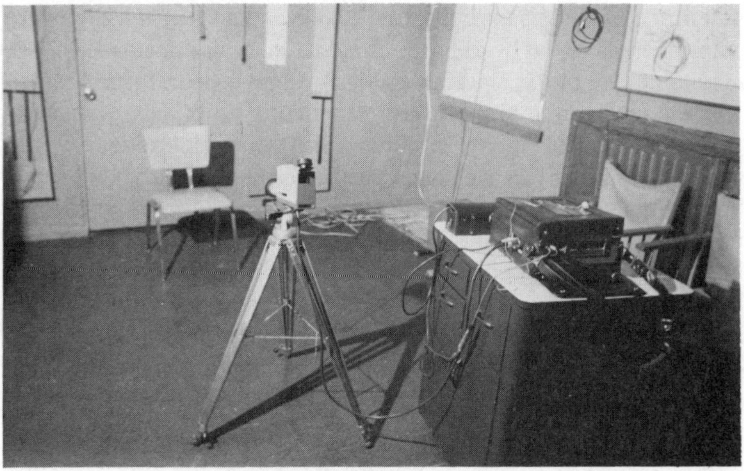

Fig. 10-1. When taping video fun and games activities such as video expressions or body video, set your recording system up in an area away from the party. This video expression setup was done in my basement. Note the roll-around metal cabinet that was converted into a video utility cabinet. The unit was purchased at a neighborhood garage sale.

so to think about what they will do or say, and then roll the tape. It works best if you have three or four people in sequence reacting to the same video expression.

During playback, you get added entertainment by comparing reactions. You can either state the expression on the tape before a sequence of reactions or you can have the group guess the expression afterwards. When getting people to appear on TV for the first time, remember that not everyone wants to be a TV star (Fig. 10-2). Never force anyone to take part. I always give my guests the option of having their performances erased.

Body Video. Body video is similar to video expressions. Have your guests come to the recording area, one at a time, and have each one select a part of his body or clothing that you can focus on with a close-up. This might require a close-up lens. Hold the close-up for 60 seconds then pull out to a full shot. After you have recordings for all your guests, play the tape back and have the group guess whose "body video" they are looking at. They will have 60 seconds to guess. Sometimes I divide the group into teams and create a live quiz show.

INSTANT REPLAY

Coaches, dance teachers (Fig. 10-3), gymnasts (Fig. 10-4),

golf pros, and tennis instructors (Fig. 10-5), just to name a few, are using instant replay to critique their students' performances. For maximum flexibility, a portable video system is best. Instant replay will test your ability to apply basic production skills and techniques.

In most cases, indoor or outdoor lighting will be adequate. Just make sure, when you are setting up the camera, not to point it at a bright light source such as an open window when you are inside or a bright horizon when you are outside. When replaying outdoors, the lighting can cause problems because a TV image is hard to see in broad daylight. If you are using a portable camera for playback, its small recessed monitor permits good outdoor viewing, but only for one or two persons. For playback before a larger group, you will need to shade a TV screen.

In most cases, sound won't be an important requirement. A possible exception is dance ; drum beats or other music is important to establish the basic timing of the performance. When music is necessary, the microphone in the camera is usually all you need. If you do use an external microphone, mount it on a solid base so that you don't pick up vibrations. And remember to use your headphones to monitor the audio during the recording session.

The camera point of view is probably the most important aspect of instant replay. You must think about camera placement

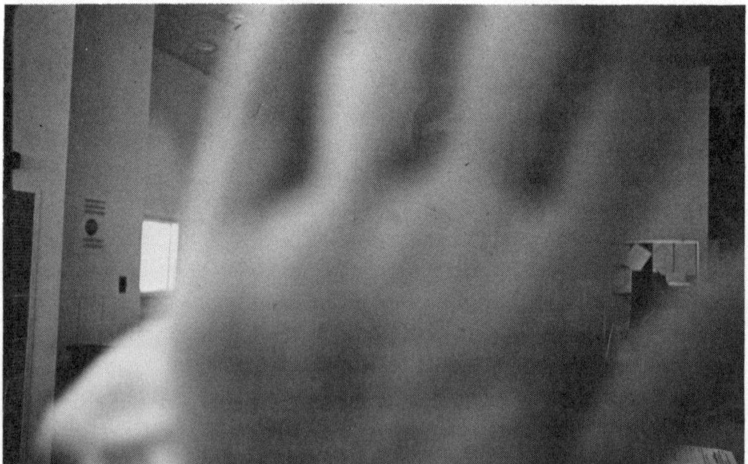

Fig. 10-2. Some people just don't like to be taped. Never force people on camera. Sometimes letting them operate the camera will make them less afraid of appearing before the camera.

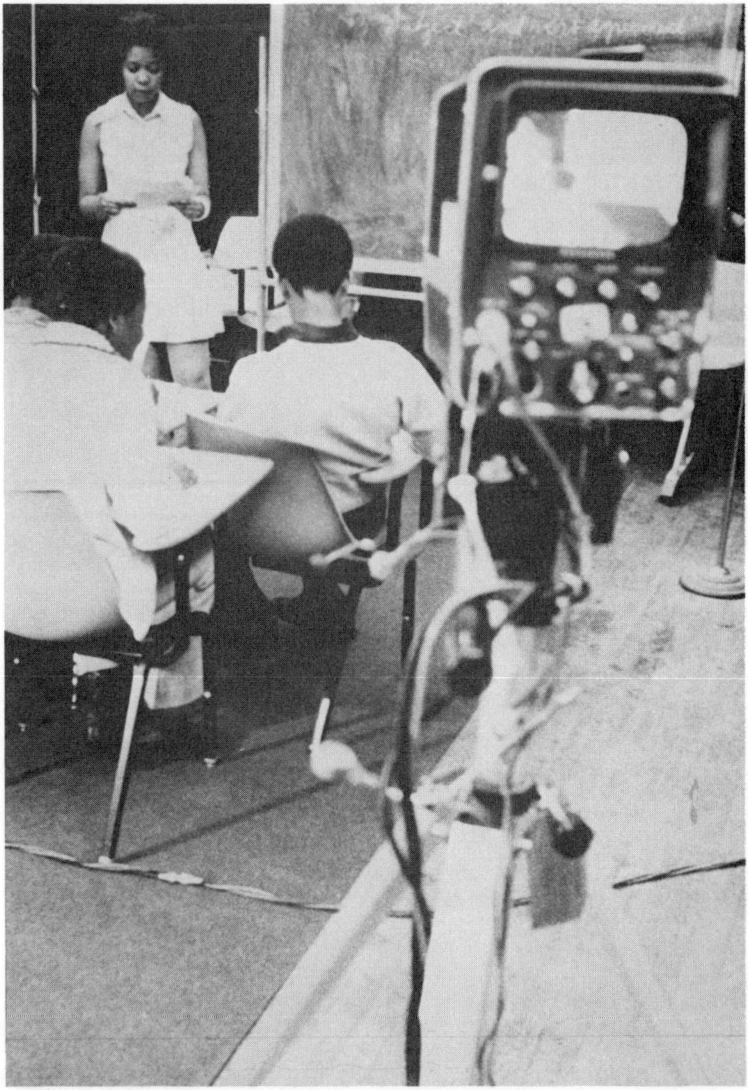

Fig. 10-3. The instant replay model can be used to critique a teacher's performance before a class.

and picture framing *before* you start recording. To select the best camera angle (high, medium, low) and the best framing (close-up, medium, or full shot), you must know exactly what you want to critique before you turn any switches. Take the time to experiment. Remember that one of the great things about this medium is that it

Fig. 10-4. Instant replay provides both athletes and performers with an excellent means of critiquing their movements.

records on reusable videotape. I've found that a zoom lens allows the most versatility in shot selection. A static or fixed shot is usually best; too much panning, tilting, and zooming can distract from the action.

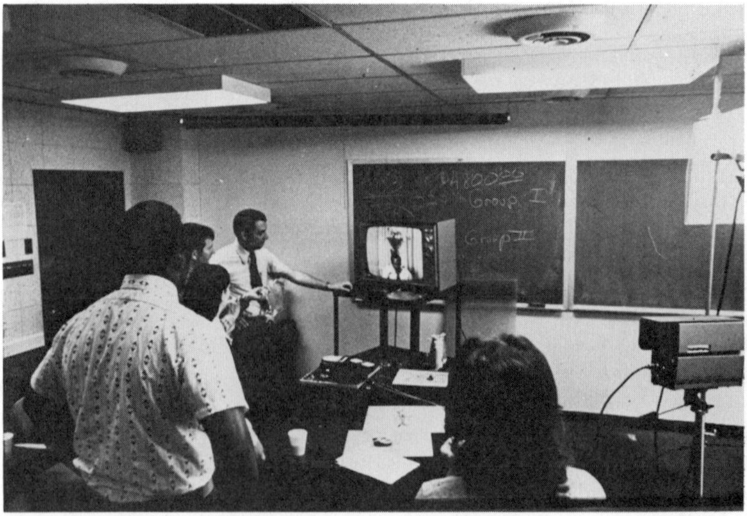

Fig. 10-5. A group of instructors reviewing their videotaped role plays. Critiquing sessions need to be structured and planned just like the taping sessions.

Here are some suggestions to assist you in planning and taping instant replay.

□ Always allow adequate time for playback. Playback might require double or triple the time needed for recording because you'll be stopping, rewinding, and starting the VCR, and perhaps doing some still frames and slow motion. This planning of your record and playback time is especially important if you are using a battery-powered unit.

□ Before leaving home, make a list of the specific skills you want to critique. This aids in the selection of camera point of view and framing and results in more constructive critiquing during playback.

□ The night before a taping session, I draw a diagram of my equipment setup. This gives me a quick checklist of the items I will need the next day. You feel ridiculous when you arrive on the tennis court only to find you have forgotten the tape. Don't laugh! It can happen to you!

□ If you are taping a group of people, you might want to play the tape back for the coach and let him make comments on the tape using audio dub. The tape can then be replayed later at each student's convenience.

□ This suggestion is what I call the golden rule of video: "Always make a test recording."

ROLE PLAY

Role play is a training activity in which the participants are taped "playing out" certain situations they experience in their day-to-day lives. After the role play has been completed, the participants and the instructor critique the performance. Video role playing is a very effective training device for teaching and improving sales skills, interview techniques, customer handling, and crisis intervention.

Most role play "scripts" consist of character descriptions rather than word-for-word dialogue. Here is an example of a role play that was used during a workshop for apartment managers:

Participant 1: Apartment manager.

Participant 2: Tenant who has occupied the apartment for a year. He has come to the manager's office to complain about the smoke in the halls and his lack of hot water. He is threatening to withhold his rent until something is done.

The person playing the apartment manager is playing his real life role. He is not told in advance of the role play what the "tenant's" problem will be. The role play allows him to "practice" the type of situation he will confront in his real job.

Most role plays involve two people and the situation is "played out" in 5 minutes or less. Most role-play activities are taped with a single-camera system. It is usually best to set up the recording system in an area that is isolated from the rest of the group, and, depending on the sensitivity of the situation, you might want to isolate the playback and critiquing as well. When the "actors" have never been on camera before, I use some type of video fun and game activity to get them used to seeing themselves on TV. I always make it clear that each "actor" has the final word as to who will see his performance.

I usually set my camera up just to the back and side of the "actor" playing the imaginary role. This will focus attention on the "actor" playing his "real" role because it is the latter person who is being critiqued. A seated situation is usually most comfortable for the actors, and it is the easiest for you to tape. For the most part, the camera can be left in a fixed position. As a rule, zooms and pans distract from the role play. Nevertheless, you might zoom in on a detail such as fidgeting hands if calling attention to body language is one of the purposes of the tape.

Getting good audio is very important. An omnidirectional microphone hung above the staging area or hidden in some flowers on the table will usually be adequate. I prefer using two lavalier microphones, one on each participant, fed into a mixer. If you have a VCR with two tracks, you can feed a microphone into each channel and then play back on the mixed mode.

Normal room lighting will usually do. I sometimes use a single bounce light to bring up the overall light level. The presence of a light also creates a more "dramatic" atmosphere that helps to raise the actors' adrenalin, producing a more natural or spontaneous role play.

ROLE PLAY MODELS

The role play model carries the role play a step further. Role play models are scripted role plays in which everyone knows exactly what they will say and do ahead of time. The production standards are higher for a role play model than they are for a simple role play. In many cases, I use professional talent (people with experience in college or community theater or in industrial broadcasting).

The purpose of the role play model is to produce a sequence

that can be used to demonstrate and teach specified skills and techniques. When the role play model is completed, it is played back to a group of trainees. After a group discussion, the instructor asks the trainees to role play the situation they just saw in the role play model. The role plays are taped, played back, and critiqued.

The key to producing a good role play model is a good script. This can be difficult when you have a contract to do a tape on a subject you know little about. I have developed an effective method of creating scripts for clients like these. First, I have the instructors sit down in front of the camera and role play what they want to be in the role play model. I tape this session using a fixed camera, available light, and two microphones (sound is very important).

Next, I make a transcript of the dialogue. I do this by first making an audio cassette copy of the videotape. Then, using a cassette recorded with a pause and back space control, I can type a word-for-word transcript. The trainers can then go over the script to make content and grammatical corrections. Next, I type the corrected transcript in a vertical script form (left-hand column contains the dialogue, right-hand column is left blank for visual directions).

I review the instructors' tape and note important visual content such as body language. Finally, I return to the script and fill in the left column with detailed visual directions. Now the script is complete!

You can use a single camera for 2-minute or 3-minute role play models. Camera point of view is very important. I frequently use an elevated angle. I set the camera up 7 or 8 feet to give the viewer a "surveillance" type of view. There is a feeling of "dropping in" on the scene. The camera shot is fixed throughout the short sequence. If facial expressions or hand gestures are important, an eye-level or low-level shot might be better because you will want to zoom in for a close-up.

Longer role play models should be shot with a two-camera system or by *double shooting*. Double shooting is a lot easier and cheaper than a two-camera shot. Let's say you are taping a dialogue between a "salesman" and a "customer." First, I would setup an over-the-shoulder shot of the "client" and tape the whole scene focusing on him. I then tape the scene a second time; this time I focus on the "salesman." Before shutting down, I get some reaction shots or listening shots. I now have two complete takes for the client to review. I combine and edit the tapes according to his instructions.

When taping a role play model, I use two lavalier microphones, and I make sure that I don't touch the mixer's level control between takes. If the participants are seated, a three-point lighting set up is usually adequate. When I change the camera, the key light becomes the back light, and the back light becomes the key.

THE VIDEO CATALOG

The video catalog is used to "display" products. I have made video catalogs for customers ranging from fashion designers to computer software manufacturers. For the most part, the catalogs are made for salesmen to use when meeting with store buyers. Obviously, it's a lot easier to sell this concept to a salesman who has a video system himself or whose clients have video systems in their stores or offices. Most of the big chain stores have video tape training networks, and someone in the training department will usually provide the gear necessary for salespeople and buyers to play back a video catalog.

My first step in producing a video catalog is to make an "ID" for each product. The ID contains the name of the product, the manufacturer, the model number, and the price. The client then brings the products to my studio where I can tape them in a controlled environment. I usually make a voice-over narrative using the clients product brochure. If the manufacturer has TV or radio spots, I space these throughout the catalog.

I recently completed a video catalog—for use in a POS (point of sale) exhibit—right in the showroom. The tape was done for a video game manufacturer and contains 60 second demonstrations of each of its games. The advantage to this is that game cartridges do not need to be opened in order to show the game graphics and concept.

THE VIDEO SCRAPBOOK

The video scrapbook is used to document family gatherings, trips, and special occasions. A popular strategy is to put a tape together using old family photographs and other mementos such as newspaper articles, wedding-cake figures, hotel menus, etc. You can develop your own dialogue or music and lay that down first, then—with insert editing—add the visuals. The video scrapbook is popular at anniversaries, birthdays, and retirement parties.

I have found the best strategy is to cover the special events such as party preparations, guests arriving, cake cutting and gift opening. Then I conduct a few select interviews either during or just after the party. The one thing I try not to do is turn the party into a major production scene. I stay as much in the background as possible. I usually take some stills or I have someone else at the party take them for me. Later, I edit the tape and insert the stills where I need transitions.

When I go on trips, I concentrate on documenting "episodes" such as the kids yelling and fighting in the backseat or a demonstration from my wife on how to ride the surf. I conduct short interviews along the way with interesting people we meet. I use the subjective camera approach to cover "tours" of interesting places. I even record some of the local TV news shows. On returning home, I edit all my screens together, and insert stills and close-ups from a road map to provide transitions from one episode to another.

In creating tapes for my video scrapbook, I use the microphone in the camera and shoot in the available light. I use audio dub quite a bit. For instance, if I videotape my wife playing tennis, later that night we'll put together a commentary to go with the sequence. I often use a special lens such as the fish-eye. Sometimes I use a special-effects lens such as the "multiple image" or "star" and filters. I find these lenses and filters especially good for creating segments to use as transitional devices or as backgrounds for titles.

SPECIAL EVENT VIDEO

Special event video refers to documenting events such as weddings, graduations, and meetings. Preproduction planning is the real key to success in this type of production. It is especially important that you visit the place where the event will take place prior to the taping. You should carefully study and diagram your setup. Contact everyone who will be directing the event and alert them to your presence. Above all, give yourself plenty of time so that you don't create a disturbance in setting up just as the guests are arriving.

Camera point of view is very important because you might not be able to move around during the actual event. This is another reason for setting up early. I try to set up my camera within 25 feet of the event. For weddings, I set up to the side and back of the minister. In taping a wedding ceremony, I have found that the tape is effective only if you can see the couple's faces. Often I set up a stationary, unmanned camera if my presence will be distracting. The minister usually doesn't object to wearing a lavalier microphone. In most cases, that is all that is needed to pick up the voices of the minister and the couple. As a backup, however, I usually place a unidirectional microphone on a stand as close as I can get it. The old "microphone in the flowerpot" trick also works well here.

When covering events such as graduations or meetings, you will often find that a public address system is being used. You can usually get a line signal out of the PA amplifier and feed it directly into your VCR's line input. Make sure you do a good test recording to check for an impedance mismatch. Also, have at least one of your microphones set up as a backup in case something goes wrong with the PA system.

Lighting can be very tricky (especially weddings). This is another reason for a preproduction visit. I ask the minister to turn on the lights that will be used during the ceremony, and I take some test pictures in that light. If it looks like I'll need more light, I ask if any more of the "house" lights can be turned on. If not, I show my client the test recording and let him decide whether to go ahead as is or to have me bring in some fill light. Special effects lenses and filters are very useful here in creating shots for use in titles or transitions. Most of the time I can get copies of stills taken during the event. I insert these in the tape either as single shots or as montages.

Reception coverage is usually done with a portable system. Once again, you must do some preproduction planning so you'll be in the right place at the right time. Make sure you know who is directing the event, and that he lets you know ahead of time when important events will occur (cutting the cake, tossing the bouquet, leaving for the honeymoon).

After the event has been covered, I edit the tape. Later, I give the client both the edited program and the original footage. One note of caution is that covering this type of event is not easy. I strongly recommend that the first few times you do it, you consider it "practice." Also, don't make the mistake of trying to do everything by yourself. Get some help. Give yourself *plenty* of time and make lots of test recordings.

THE TRAINING TAPE

The following are some guidelines that will help you in producing training tapes for small manufacturers, schools, county and state government agencies, and other organizations. The training tape has the greatest potential for revenue. It also has the greatest potential for problems. In making a training tape, it is of utmost importance to define objectives and to assign specific roles and responsibilities.

First, I have a conference with the client. I ask him to tell me, in a few brief statements, what he hopes the tape will accomplish. Next, I ask the client to designate a "yes-no person" in the organization who will have the authority to approve each step in the production process. The client must understand that time is money, and that once a procedure is approved by the yes-no person, it cannot be redone (unless, of course, the client wants to pay for it). I tell the client that these are the items that must be approved:

The Treatment. A scene-by-scene description of the action to take place along with what needs to be said.

The Script. A detailed outline that lists the dialogue visual action separately.

Location Shoot. The shooting of source material for later editing.

First Edit. A rough edit that is used to "test" transitional devices and continuity, timing of the shots, etc.

The Final Edit. The production of the finished product.

Depending on the complexity of the tape, the production will involve one or several people in each of the following categories: subject matter specialist, script development, graphics, and equipment.

Subject Matter Specialist

A subject matter specialist is the authority on the tape's content or that part of the tape that involves his special knowledge. You might have several subject matter specialists if your tape deals with several content areas. Beware! One subject matter specialist per content area is enough. Experts tend to disagree. Isolate the subject matter specialists from each other (in time or space) if you want to keep your production on schedule. Remember, too, that the perception of each subject matter specialist is obscured, to some degree, by the very fact of his specialization. Your yes-no man should be able to give you a good "anchor" to pull yourself back to earth when the custodians or the food service staff have convinced you that theirs is the most important job in the organization. Don't laugh. It has happened!

Your subject matter specialist must also be reliable and he must be available, when you need him, with relevant, objective information.

Script Development

The script writer will take the objectives, the data provided by the subject matter specialist(s), and any other relevant material and turn it all into colloquial English in a logical sequence. He can use humor, juxtaposition, flashbacks, and other devices to heighten the audio and visual appeal of the production. Just as there is no one way to make a videotape, there is no one way to write a script.

The script writer almost always begins with a rough draft (after discussing the concept of the tape with all involved). Successive drafts, coordinated with others in the production, should enable him to produce a final draft that meets the approval of the yes-no man.

The script writer should not be responsible for operational or policy decisions. You and the yes-no man should deal with disagreements among subject matter specialists or squabbles between the production and graphics people. This allows the script writer to concentrate on the writing.

Graphics

The graphics person is responsible for the visual inputs to the script such as charts, graphs, slides, titles, cue cards, and film clips. All graphics must be in TV format (three units high by four units wide—with at least two units of bleed space on all sides—to allow for framing). If possible, reduce all graphics to the same size and medium (this is not absolutely necessary). The cardinal rule for TV graphics is that they be legible and understandable on the TV screen in the duplicated distribution copy.

Equipment and Crew

Specific effects and points of view—such as an overhead traveling shot from a crane or a low-angle shot as an auto runs over the camera position—will require careful coordination of equipment and crew. Remember that the sophistication of available equipment and crew will set finite limits on the type and variety of shots, editing, and production effects. Script and production capability must be well coordinated in the preproduction process to prevent your tape from becoming "amateur night at the studio."

Production Coordinator

Production coordinator is your role. You must coordinate the activities of all the people involved. You must also set up and maintain a production schedule. You might well find yourself playing more than one role; just don't try to do too much at once. A preproduction planning form is something I have found most helpful.

It is structured to get all the essential information quickly and concisely.

Preproduction Planner

Your client should provide as much information as possible in responding to the following questions.

☐ State the goal(s) of this tape.

☐ Who is the specific audience?

☐ After watching the tape, what will the audience be able to do in terms of specific observable behavior (measurable change)?

☐ What level of performance is required. Will 90 percent of all people watching the tape be able to perform all the instructional objectives?

☐ Check which strategy would work best for accomplishing the instructional objectives:

—Videotape of different scenes with narration over finished tape.
—Videotape interviews with subject matter experts.
—Dramatization of scenes rather than actual situation using employees or professional actors.
—Interview with authority with supporting graphics and videotape inserts.
—Panel discussion with several authorities.
—Other.

☐ Outline the specific points to be made in the tape in the order of their occurrence.

☐ What support materials could be included in this tape to make it more effective (charts, diagrams, slides, film, videotape, audio tape, pictures, etc.)?

☐ Specify a sample of the target audience that can review the final script.

☐ Specify a "yes-no" person who has overall budget authorization and the final approval of all content.

☐ Indicate any policy problems, clearances, and other concerns that should be resolved before final approval of the project.

Remember that this is only one way to make a videotape. It's not the only way. Experience teaches things about your clients. Some of the things I've learned the hard way are:

☐ A client might *think* he knows precisely what he wants. He probably doesn't. Unless you help him clarify his preferences in advance, you will suffer for it.

☐ Even in the most rigidly structured organizations, such as police and fire departments and the federal government, getting a firm decision on what to tape (and what not to) in *advance* will be difficult. That's why we insist on a yes-no man. He won't be able to help you 100 percent of the time, but even 50 percent of the time is better than nothing.

☐ The public will subconsciously compare your efforts to broadcast TV because it is their only frame of reference. Therefore you *must* strive to match their technical quality. It can be done, even with the simplest half-inch VTR system, if you take the pains. But it doesn't come automatically.

☐ Don't become the subject matter specialist. You've got enough to do as producer. And don't let arguments among subject matter specialists (and these are inevitable) interfere with the progress of the tape. Use your "yes-no" person.

☐ Don't allow the apparent complexity of this process to overwhelm you. It is an effort to help you gain control of the production process so you can communicate. Without the active exercise of control by one person, the innumerable variables of the production process run amuck. Before you can say "Marshall McLuhan," the medium is getting in the way of the message.

Just remember that if someone showed you a schematic flowchart of the operation of your automobile, you'd probably be totally intimidated. Yet you might jump in that same car and head to California without a second thought. So don't get hung up on the technicalities and mechanics of the system. Learn to control it so that you can use your creative power to say something worthwhile. Don't waste dealing with a lot of unnecessary and time-consuming minicrises.

VIDEO STAR

Everyone wants to be a TV star! Well, now they can be. Video Star is a production strategy I use in marketing the video portfolio concept. There is something special about seeing a person on TV if he or she is presented in the right perspective. I got started with Video Star when a model asked me to tape an audition for her by editing together a number of TV spots she had done. When I found that she also had a portfolio of still photographs, I suggested that we combine the stills with the TV samples. I showed her how I could lay down music and overlay it with a montage of her stills. I also suggested a brief on-camera introduction in which she could give prospective clients a brief overview of her background. Showbiz!

She liked the idea and the price at $200.

The Video Star strategy can be modified for just about anyone. I have used it to "present" executives, dancers, comedians, and singers. The video format is especially good to use with people who have an assortment of "demo" materials such as photographs, audio tapes, performance tapes. The overall length of the presentation is less than 10 minutes. I divide the time into a number of segments featuring "live" action as well as short transitional montages of stills. This is an exciting application for small video professionals to pursue.

LEGAL VIDEO

Video is being used more and more to supplement the legal process. Video depositions and wills were first introduced and used as court evidence in the late '70s. More recently, video inventories and "day-in-the-life-of" documentation tapes have been introduced as evidence to support a client's case. Legal video doesn't require an elaborate production system. It is well suited to the video user with a single-camera portable system and at least one editing desk. What follows are short descriptive outlines of several legal video production strategies. Before launching into a legal video service, check with the local, state, and federal courts to see what is and is not permitted.

VIDEO DEPOSITION

The video deposition offers a means of obtaining testimony, with opportunity for questions and objections by the opposing councel. The testimony can then be stored until it is time for the trial. This approach helps eliminate delayed court trials due to unavailable witnesses. It also gives defendants and plaintiffs a more economical means of getting testimony from expert witnesses who previously had to be flown in, fed, and housed for court appearances.

Videotaping a deposition is not very difficult. The witness is seated at a table (Fig. 10-6 and 10-7) with the lawyer for the plaintiff on one side and the lawyer for the defendant on the other. In an acoustically sound room (a lawyer's library or conference room), you can generally use a good omnidirectional microphone mounted on a stand in front of the witness. I often set up at least one bounce light to bring up the overall room light. My shots vary from shoulder shots of the witness to two shots of the witness, and whoever is asking the questions.

Most courts now require a record of the running time to appear on the screen. To do this, you will have to buy a "time-date" generator. This is hooked to your system between the camera and the VCR. Therefore the video signal in your camera is looped through the generator. Actually, I advise using a time-date generator even if it's not required because it gives you an effective means of logging and indexing your tape. When objections are raised, you need to note exactly where they occurred.

If an objection is raised, the attorney doesn't stop ; he proceeds with the next question, and the witness answers it. Later, the judge is given a tape log that indicates the location of the objections. He reviews the tape or a transcript and makes a ruling on the objection. If he overrules the objection, the tape is fine as is. If the objection is sustained, you can do one of two things. You can edit out the answer or, when the tape is played back in court, you can push the "fast forward" button when you come to the sustained objection.

You need to know in advance if the witness will have to move around. Witnesses will often have charts and models they use in their testimony. If you give them lavalier microphones you will need a mixer.

At the beginning of the deposition tape, I have the attorney introduce first himself and the witness. The witness is then sworn

Fig. 10-6. An ordinary office has been changed into a taping room used by a legal firm to do depositions and wills.

Fig. 10-7. Notice the addition of drapes for sound proofing and added fluorescents to elevate the normal room light.

in on camera and the testimony begins. The first few times you do this I would advise having a court reporter present to provide a backup, written transcript.

Civil cases are well suited for video deposition. Criminal cases, on the other hand, require that the defendant, as well as the judge, be present during depositions. This can develop into a major production—the type you might want to avoid. One note of caution is that lawyers love to be "in control." If you display the least bit of confusion, they will jump in and offer "expert" advice. Stay calm, stop what you're doing, collect yourself, and begin again.

For more details on video depositions, write to:

The Federal Judicial Center
1520 H Street NW
Washington, DC 20005

and ask them to send you a copy of their booklet entitled, *Guidelines for Videotaping Depositions in the Federal Court.*

THE VIDEO WILL

The video will was developed primarily to demonstrate a person's competency at the time of writing the will. It is usually accompanied by a written document. The lawyer opens by introducing the client and any witnesses present. He then swears in the client, and the client reads the will. The client may include such comments as, "Dorothy, you and John always liked that old car of mine, so I am leaving my 1929 Bentley to you."

Production Setup. If the person reading the will is in good health, I light the room with either a three- or a two-point lighting system. If he is ill, you must use your own discretion about how much light to use. I use lavalier mics (one on the lawyer and another on the client).

DAY-IN-THE-LIFE-OF VIDEO

The day-in-the-life-of video technique was developed to show a judge or jury what it is like for a disabled person to perform day-to-day activities. It is used to support cases in which an individual is suing a company, holding it responsible for his disabling injury. The tape usually starts off in the morning with the client getting out of bed, going to the bathroom, getting dressed, preparing breakfast, going to work, etc.

Production setup. A portable system is best here, and an advance visit to the production site is an absolute necessity. I generally use a single light either mounted or hand held right above the camera where I can use it to create either bounce light or direct lighting. I normally use the microphone in the camera. This type of production can be shot "inside the camera." During my preproduction visit, I have the client go through all the activities he will be performing on the tape. I have the attorney present to answer questions about camera angles, close-ups, and other movements that could be objected to by the opposing party. I make diagrams of the areas where I will be taping. On returning to my studio, I plot out my camera positions and angles and my equipment setup.

During the actual taping session, the attorney is present again. After shooting a sequence—such as getting out of bed—I stop the tape, rewind it and play it back for the attorney. I then set up for the next sequence (assembling it onto the tape following the last sequence). As with other types of legal video, I have the lawyer introduce the client at the opening. After that, the only audio on the tape is what occurs as the client goes through his daily routines.

Playback. In most cases, I play back this type of presentation over a multimonitor setup. I have used large-screen projection in the courtroom, but you often encounter problems with lighting and viewing angles, particularly if a jury is involved. The lawyer gives a running commentary, and stops and starts the VCR at appropriate points. I do recommend monitors over receivers. Receivers encounter a lot of rf interference in the courtroom caused by the use of two-way radios for security reasons.

ENVIRONMENTAL VIEW

I developed the environmental view technique to provide a judge or a jury with an overview of the environment in which a crime occurred. I used it for the first time in the early '70s while videotaping a magistrate's hearing in the Macon, Georgia State Prison. A record of the hearing was to be made on videotape for the judge, who would make the final ruling based on the arguments presented in the tape and the recommendations of the magistrate.

A key factor in the hearing was an incident that occurred in the maximum security cell block. I suggested a "tour" of the cell block with the video camera. We all set out together—the guards, the magistrate, the attorneys for the defense and the prosecution, and myself. What resulted was a "walking tour" of the cell block with a commentary by the attorneys. I used the microphone in the camera to record the commentary. I wore headphones so I could better hear what they were saying in order to focus the camera on the areas they described.

More recently, I was involved in a case where the attorney wanted to show the judge and jury the environment in which a car chase had taken place. I suggested we combine the actual radio transmission from the squad car with video footage. I copied the radio transmission onto a cassette, and then I got into the squad car with the policeman involved in the chase. I asked him to retrace his route as we replayed the radio transmission. I videotaped the drive and synchronized it with the audio transmission. On returning to my studio, I first laid down the audio transmission and then inserted the video. This technique will be used more and more in cases where either plaintiffs or defendants want the jury to see where a crime or accident occurred.

BEDROOM DOCUMENTARIES

No discussion of production models would be complete without mentioning the category I call bedroom documentaries. Here are a series of activities guaranteed to launch you into a totally new video experience!

VCR Foreplay

My first experience in using video in the bedroom involved an exercise I now refer to as VCR foreplay. This is not a documentation of actual sexual foreplay. It is a video technique used to acquaint the partners with a VCR. It's a mild-mannered way to relax participants, and to lessen the initial panic or shock for people seeing themselves on a television monitor for the first time.

VCR foreplay usually involves taping one partner dancing, reading provocative literature aloud and to the camera, or even audio dubbing over a copy of one of those silent "blue flicks."

Try playing with some lighting effects, if you're taping someone dancing, or use those special-effects lenses. A strobe light or revolving mirrored ball can create interesting effects. You might also want to project a slide or moving picture on a dancing figure. Costuming the dancer can add to the fun.

Sexual Surveillance

Sexual surveillance is just what the name implies. Prop the camera on a tripod placed at the foot or the side of the bed. Turn on the VCR and do it! Audio is optional here. You can prop a microphone up near the bed and play the whole thing straight or you can add your own comments and/or sound effects later, using audio dub.

Hand Holding

Hand holding is the name of the technique used to get the camera off of the tripod and into the hands of the participants. One modification needs to be made to the camera; replace the normal zoom lens with a wide-angle lens. The wide angle makes the camera far easier to maneuver and frees the operator from having to make numerous lens adjustments. The wide angle has a far greater depth-of-field than the zoom, so less attention need be given to focusing.

Hand holding puts a new and exciting perspective on the art of lovemaking. It's important that both participants take part in the camera work. It is extremely interesting to "see" yourself under these unusual circumstances. By observing what your partner allows the camera to record, you gain in sight into what really turns your partner on. Almost all shots using hand holding are, necessarily, close-ups, and this results in a totally new understanding and view of areas of the body. Imagine being surrounded by mirrors that can hold images and play them back at a later time. That's the only way to describe it!

There is a warning attached to the hand holding technique. This exercise otherwise eliminates the use of one hand and often an entire arm, and sometimes it creates a need for new positions that you and your partner might not be comfortable with. You can always study positions outlined in the *Kama Sutra*. Remember that practice makes perfect.

One additional key to hand holding is to have the TV monitor positioned so that the person doing the hand holding can refer to the screen from time to time. Make sure you are getting what you want in the picture. This periodic referral to the screen has its own side effects. It can be very erotic and stimulating. So be careful trying this technique. Someone who's never seen themselves on TV might feel a little overexposed at first. This is one reason why it's advisable to begin boudoir video with VCR foreplay or sexual surveillance. Work up to hand holding.

VCR Fantasy Skits

VCR fantasy skits are very involving approaches to bedroom documentaries. The general format places partners in long-range projects that, in essence, combine talents and skills developed in the three other exercises. First, the partners discuss various sexual fantasies they would like to see dramatized for television.

Once a fantasy has been selected, the partners can begin to develop a storyboard or shooting script. This means making a list of the action shots that are to take place. If you have a talent for drawing, you can combine a written description with thumbnail sketches depicting various shots and camera angles to be employed.

Start off simply. Try to make the preliminary fantasies easy to depict. Keep your production time short; a 10-minute fantasy dramatization is long enough for beginners.

The End

Well that's about it for now. Actually this is not the end but the beginning. The beginning of what I hope will be for you many hours of enjoyment and profit in using your video system. The technology will change. The gear will get shinier, the prices will get lower (we hope) and your use of this amazing media will grow and grow. So for now, good luck and good taping!

THE VCR

Madly I dash about getting ready to tape,
Setting up my VCR pictures and words I'll try to take,
And make something out of the electronic shapes
That momentarily fix themselves upon my face.
Pointing the camera, focusing the lens,
Go back, pause, search, fast forward
The visions flash by in double, triple
The tube flashes light, dark, stop, play.
Instant instance!
Look into the viewfinder and see the picture glow
Reach out with hand, zooming in...out
Madly I dash about, getting ready to tape
Setting up my VCR, lugging it in from the car.

Recommended Reading

The following list contains the names of trade magazines, books, and newsletters that are of interest to video users.

Educational and Industrial TV
C.S. Tepfer Publishing Co.
51 Sugar Hollow Road
Danbury, CT 06810
Videoplay
C.S. Tepfer Publishing Co.
51 Sugar Hollow Road
Danbury, CT 06810
Home Video
Media Horizons
475 Park Ave. So.
New York, NY 10016
Modern Photography
ABC Leisure Magazines Inc.
825-7th Ave.
New York, NY 10019
Videography
Media Horizons, Inc.
475 Park Ave. So.
New York, NY 10016

Video User
Knowledge Industry Publications
701 Westchester Ave.
White Plains, NY 10604
Video
Reese Publishing Co. Inc.
235 Park Ave. So.
New York, NY 10003
Video Pro
Video Pro Publishing
350 E. 81 Street
New York, NY 10028
Video Review
Viare Publishing
350 East 81 Street
New York, NY 10028
Home Entertainment
Home Entertainment, Inc.
220 Westbury Ave.
New York, NY 11514
Video Business
CES Publishing Co. Inc.
135 West 50 Street
New York, NY 10020
Audio Visual Directions
Montage Publishing
25550 Hawthorne Blvd.
Suite 314
Torrance, CA 90505
Video Store
Hester Communications
1700 E. Dyer Road
Suite 250
Santa Ana, CA 92705
ETV Newsletter
Videoplay Report
C.S. Tepfer Publishing Company
51 Sugar Hollow Road
Danbury, CT 06810

Broughton, Irv *The Art of Interviewing for Television, Radio and Film*. TAB book No. 1125.

Clifford, Martin *Microphones*. TAB book No. 1475.
Efrein, Joel *Cablecasting Production Handbook*. TAB book No. 768.
Efrein, Joel *Video Tape Production & Communication Techniques*. TAB book No. 541
Ellmore, R. Terry *Broadcasting Law & Regulation*. TAB book No. 1450.
Everest, F. Alton *Handbook of Multichannel Recording*. TAB book No. 781.
Everest, F. Alton *How To Build a Small Budget Recording Studio From Scratch . . . with 12 tested designs*. TAB book No. 1166.
Goodman, Robert L. *Maintaining and Repairing Videocassette Recorders*. TAB book No. 1503.
Hood, Edmond W. *Practical Handbook of Stage Lighting and Sound.* TAB book No. 1083.
Ingram, Dave *Video Electronics Technology*. TAB book No. 1474.
Jorgensen, Finn *The Complete Handbook of Magnetic Recording*. TAB book No. 1059.
Knecht, Kenneth B. *Designing & Maintaining the CATV & Small TV Studio–2nd Edition*. TAB book No. 815.
Kybett, Harry *The Complete Handbook of Videocassette Recorders–2nd Edition*. TAB book No. 1211.
Paiva, Bob *The Program Director's Handbook*. TAB book No. 1363.
Pasternak, Bill *Video Cassette Recorders: Buying, Using and Maintaining*. TAB book No. 1490.
Skiles, Marlin *Music Scoring For TV & Motion Pictures* TAB book No. 779.
Traister, Robert J. *Make Your Own Professional Home Video Recordings*. TAB book No. 1433.
Wardwell, Douglas O. *Television Production Handbook*. TAB book No. 1161.

Glossary

AAE—Automatic assemble editing.

A-B rolls—Videotape with the same information recorded on both rolls but with different camera views.

ac—Alternating current. Standard 120-V, 60-cycle (Hz) USA household current. VTRs and cameras often use 60 Hz as a horizontal sync reference.

adjacent channel—Any of the two TV channels are considered adjacent when their video carriers, either off air or on a cable system, are 6 MHz apart. FM signals on a cable system, two channels apart are adjacent when their carriers are 400 to 600 kHz apart.

AEF—Automatic edit function. An automatic assemble editing feature found on JVC and Panasonic ¾″ portable VCRs.

AFT—Automatic fine tuning.

allocations—The assignments of frequencies by the FCC for various communications uses (television, radio, land-mobile, defense, microwave, etc.) to achieve a fair division of the available spectrum and to minimize interference among users.

alternative TV—Non-conventional TV programs and production processes.

amplifier—A device used to boost the strength of an electronic signal. Amplifiers are spaced at intervals throughout a cable system to rebuild the strength of television signals that weaken as they pass through the cable network.

angle shot—A shot taken from a position either above or below standard tripod height.

antenna—A device used to transmit or receive broadcast signals.

aperture—The lens opening, f-stop, or iris that determines the amount of light that enters the camera.

aspect ratio—The ratio of the height to the width of a television picture (which is 3:4).

assemble editing—A new video or audio sequence that is consecutively added to a previously edited scene.

audio—The aural components of a television signal; also, aural broadcasting such as AM and FM radio.

audio cue—A sound that is used to "tag" or denote an upcoming production event.

audio dub—The recording of sound only (without disturbing the picture).

audio head—The unit of an audio or video recorder that receives the sound signal and transfers it to the tape, and/or receives the signal from the tape for reproduction.

audio-in—The input jack that delivers an audio signal to a piece of equipment.

audio limiter—An electronic circuit that adjusts or limits the audio or video intensity to a preset level.

audio mixer—An electronic circuit that receives audio signals from various sources, and enables the audio technician to balance their volumes to form one composite sound signal.

audio out—The output jack that delivers an audio signal from a piece of equipment.

audio track(s)—The portion of a videotape on which the sound is recorded. In the ¾-inch cassette format, there are two audio tracks.

automatic editing controller—An electronic programmer used in conjunction with editing VCRs to facilitate editing.

automatic gain control—An automatic circuit that controls the audio or video intensity during recording.

auto search control—An optical remote control unit used with certain industrial VCRs.

auxiliary—Line level (low impedance) audio input. Cannot be used as a microphone input.

available light—The amount of light normally present in the environment.

azimuth—The angle of a particular recording head.

background—A term that usually refers to the background of the setting itself, but which can also refer to background music or background sound effects (either of which are played at a level that is subdued and below the level of the dialogue of the performers).

back light—Illumination of the subject from behind to produce a highlight along its pictured edge.

back porch—That portion of the composite picture signal that lies between the trailing edge of the horizontal sync pulse and the trailing edge of the corresponding blanking pulse.

backspace—A positive, reverse drive for tape or film to allow arrival at an established preroll point (park point).

bandwidth—A measure of spectrum (frequency) use or capacity. For instance, a voice transmission by telephone requires a bandwidth of about 3000 cycles per second (3 kHz). A TV channel occupies a bandwidth of 6 million cycles per second (6 MHz). Cable system bandwidth occupies 50 to 300 MHz on the electromagnetic spectrum.

barn door—Shutters or flaps (usually two or four) that are attached to the front of the luminaire in order to control the shape of the light beam.

base light—Uniform, diffuse illumination used to establish a sufficient ambient level of light for quality type television and film pickups at desired lens aperture. Customary base light levels: for standard three-tube Plumbicon color cameras, 200 ft-c (foot-candles)-400 ft-c; for image-orthicon monochrome cameras, 75 ft-c-100 ft-c; for portable monochrome vidicon cameras, 100 ft-c-300 ft-c, with 200 ft-c-250 ft-c the norm.

basic set—The television setting as it exists without props or other production material on it.

batten—Horizontal pipe on which luminaires or scenery can be hung.

beam—An electron gun in a camera or picture tube emits a flow of electrons called a beam. The beam creates the scanning lines as it sweeps the tube's target area.

Betacord—Sanyo home VCR.

Betaformat—Toshiba home VCR.

Betamax—Sony 1-hour, ½-inch videocassette recorder.

Beta-2—Sony 2-hour, ½-inch videocassette recorder.

Betavision—Sears home VCR.

bias lighting—An electronic boosting of low light levels that enter the camera so that the pickup tubes can operate relatively noise free.

black—The recording of synchronization pulses only.

B&W—Black-and-white.

black clipping—A video control circuit—found in cameras and VTRs—that regulates and contains the black level of the video signal so that it does not disturb or appear in the sync portion of the signal.

black level—The bottom level of the picture signal, below which are the sync, blanking, and other control signals that do not appear as picture information.

blanking—Related to the scanning process of a camera or receiver (whereby the retrace lines are blanked out).

B-load—A type of cassette loading design used in Betamax and Beta VCRs.

blocking—The process used by the director in planning the physical movement of people and objects on the set. Placement of people is sometimes simplified by marking (with chalk or tape) certain prearranged positions on the studio floor. The participant can then "take his mark."

BNC—A professional-type bayonet video connector.

board—The control board on which are located video (and often audio) controls for cameras, recorder, and microphones.

boom—A long piece of metal piping (or substitute) used to extend a microphone or a light over the head of a subject while remaining out of camera view.

boost—To turn up ; to increase in volume.

break—To take time out, as a rest ; a director's command to a cameraman, as in "Camera 2, break for a two shot."

breezeway—In NTSC color, that portion of the back porch between the trailing edge of the horizontal sync pulse and the start of the color burst.

brightness—The intensity of objects (highlights) on a TV screen.

broad—A floodlight with a broadside, panlike reflector.

broadband—A general term used to describe wide bandwidth equipment or systems that can carry a large proportion of the electromagnetic spectrum. A broadband communications system can accommodate all broadcast and many other services.

broadcaster's service area—Geographical area encompassed by a station's signal.

broadcasting—Transmitting electromagnetic signals in a multidirectional pattern over the air.

broadcast TV—A type of programming and bureaucracy that produces TV programs specifically for over-the-air broadcasting.

bulk eraser—A device that operates on standard 110-volt ac current and creates a strong magnetic field. It is used to erase TV magnetic tape so that it can be reused.

burn-in—An image that persists in a fixed position on a camera signal after the camera has been shifted to another scene. Burn-ins are usually caused by aiming a camera at a bright light source, and can result in permanent damage to the camera.

burst flag—A pulse produced by a color sync generator ; when present, it causes the signaling color camera to produce a burst signal.

burst signal—Color burst. A set of high frequency (3.5 MHz) pulses at the beginning of each line which determines the phase of the color signal.

business—Activity and movement (other than spoken words) to make the performer appear relaxed, lifelike, and believable.

busy—A set is called "busy" when it has too many confusing shapes, details, or other physical elements in it. Such a condition detracts from the performers and their message.

cable casting—To originate programming over a cable system. Includes public access as well as local origination programming.

cable TV—Previously called Community Antenna Television (CATV). A communications system which distributes broadcast programs and original programs service by coaxial wire.

cameo—A lighting situation in TV production when only the foreground is illuminated. The background is black and unlighted.

camera—The "eye" of the video system, converting reflected light to electronic signals.

capstan servo—A system installed in a VTR to control the tape drive by using internal or external reference.

cardioid—A microphone pickup pattern. Sounds coming from directly in front are picked up best. Reception gradually falls off toward the rear of the mike in a heart-shaped pattern.

carriage—A cable system's procedure of carrying the signals of television stations to its various channels. FCC rules determine which signals cable systems must or may carry.

cassette—A self-contained package of reel-to-reel blank or recorded film, videotape, or electronically embossed vinyl tape that is continuous and self-rewinding. Similar to a cartridge but of slightly different design.

C-clamp—A metal clamp that lighting instruments are attached to at the lighting battens.

center—This is a director's term given to a cameraman so that he will frame a subject in the center of the picture. The command might also be to "center-up."

certificate of compliance—The approval of the FCC that must be obtained before a cable system can carry television broadcast signals.

channel—In television, a single path or section of the spectrum 6 MHz wide that carries a television signal.

channel capacity—The maximum number of 6 MHz channels that can be simultaneously carried on a CATV system.

character generator—A device for the electronic display of letters on the screen.

chroma control—Control for adjusting color saturation in a reproduced video picture.

chroma key—The electronic introduction of a color background into a scene. Unlike black and white keying, color is present and color values can be adjusted by the operator of the keying unit.

chromaticity—A subjective evaluation of hue and saturation.

chrominance—Chroma. The hue and saturation of an object as differentiated from the brightness value (luminance) of that object.

chrominance channel—In a color television system, any path that is intended to carry the chrominance signal.

chrominance channel bandwidth—The bandwidth of the path intended to carry the chrominance signal.

chrominance signal—That portion of the NTSC color television signal that contains the color information without the chrominance signal. The received TV picture would be in black and white.

cinching—Longitudinal slippage between layers of tape as it is being rewound. It can cause damage to the tape. Take-up pressure must be carefully regulated.

closed circuit—A system in which television signals are distributed from originating equipment directly to TV monitors. The distribution system may employ coaxial cable, microwave relays, or telephone lines.

coaxial cable—A specially designed cable used to carry one or more channels of TV signals. It makes possible the transmission of high video frequencies with very low loss of power, and with little or no interference.

cochannel—Any two or more TV signals are considered cochannel when their video carriers, either off air or after conversion by CATV equipment, occupy the same TV channel.

color—The characteristics of light other than spatial temporal inhomogeneities.

color background generator—An electronic circuit used in chroma keying to produce a solid color background of any desired hue and saturation.

color bars—A series of bars—each a different, precise color—used as a test pattern for adjusting color video equipment.

color breakup—Any fleeting and partial separation of color picture into its "display primary" components caused by a rapid change in the condition of viewing.

color burst—See burst signal.

color camera—A video camera capable of changing both the brightness values, (luminance), and the color values (hue and saturation, expressed as chrominance), of a scene into a series of electronic pulses.

color contamination—An error of color rendition due to incomplete separation of paths carrying different color components of the picture.

color dissector tube—Coloring tube. Chrominance tube. A cathode-ray tube designed to separate a scene's hue and saturation values into their R, B, G components for electronic encoding as part of the color video signal.

colorizer—Electronic circuitry used to generate a chrominance signal in relation to the gray values of a black and white video signal. Each graduation of gray from black to white is assigned a color value. The result is an artificially, and often inaccurately, colored picture.

color key cameras—A black-and-white camera used to super graphics over a picture.

color killer circuit—An electronic circuit used in a VTR to suppress the 3.58 MHz color carrier frequency when a black-and-white tape is being shown. The same circuit in a black and white VTR used to suppress the color carrier frequency when a color tape is being played back in black and white. Without a color killer, the color signal would appear in the displayed black-and-white picture as random noise.

color phase—The proper timing relationship within a color signal. Color is considered to be in phase when the hue is reproduced correctly on the screen.

color picture signal—The electrical signal that represents complete color picture information, excluding all synchronizing signals.

color picture tube—A cathode-ray tube, the screen-end of which is capable of glowing with the three primary colors. Its cathodes produce three electronic beams and its roster area is coated with three different types of phosphor.

colorplexer—Encoder. Electronic circuitry that processes three separate color signals—red, blue and green—coming from the pickup tubes into one composite encoded color video signal.

color subcarrier—In NTSC color, the 3.58 MHz carrier whose modulation sidebands are added to the monochrome signal to convey color information.

color sync—A control signal necessary for the operation of color cameras, SEGs, and monitors. Consists of a 3.58-MHz burst (that sets the color phase and the placement before each line is scanned), and a 3.58-MHz color subcarrier.

color tape—(a) A tape on which color television, both audio and video, has been recorded. (b) a misnomer for videotape. Any videotape has the capacity to record either black and white or color, there being no difference physically in videotape.

color temperature—The temperature of a blackbody that generates light with the closest visual color match to the source being specified.

color transmission—The transmission of a signal that represents both the brightness values and the color values in a picture.

common carrier—Any point-to-point communications relay service available to the general public at nondiscriminatory rates. The carrier cannot control message content.

compatibility—That property of a color TV system that permits substantially normal monochrome reception of the transmitted signal by typically unaltered monochrome receivers.

composite color signal—The color picture signal plus blanking and all sync signals.

composite color sync—The signal comprising all the sync signals necessary for proper operation of a color receiver. This includes the deflection sync signals to which the color sync signal is added in the proper time relationship.

contrast—The difference in brightness among various elements of a TV picture.

control panel—The master location in the control room where various selector switches for camera control are located; also may contain controls for other audio and video functions.

control room—The room in which the director and technical personnel control the production activity of a studio. They select the camera to be aired and also shade the picture and perform other electronic functions.

converter—A device used to convert non VHF television signals into standard VHF channels. Cable systems often install converters where more than 12 channels are relayed on a single cable.

Converters also protect signals from the interference of strong local signals.

cookie—Any cutout pattern that, when placed in front of a spotlight, produces a shadow pattern. The cookie—usually made from a thin, cutout metal sheet—is inserted into a pattern projector or ellipsoidal spot.

corner resolution—It is inherent in a television system that the center of the picture is always sharper than the corners. The capability of an electronic system to resolve detail in these corners is called corner resolution.

cover shot—Same as establishing shot. A view of the entire set or viewing area to acquaint the viewer with the overall background.

crop—A type of camera framing that excludes some of the subject matter from the shot.

crossfade—A type of transition that involves a very fast fade to black and a fade back into the next scene. This term is more often used to signify the crossing of two sound channels (with a brief separation of silence).

cross ownership—Ownership of two or more kinds of communications outlets by the same individual or business. The FCC prohibits television stations and telephone companies from owning cable systems in their service areas. Television networks are prohibited from owning cable systems anywhere in the U.S.

cross shot—A condition in which two cameras are shooting from opposite sides of the set. First one camera is taken and then the other.

CRT—An abbreviation for cathode-ray tube. It is the picture tube of a television system.

cue—The signal, either vocal or made visually, for action to begin.

cue card—A card placed on the set where the performer can see it easily. It contains script information, lyrics of songs, or other information he might have trouble remembering. Cue cards are also facetiously called "idiot cards."

cut—When used as a command by the director during production, it means to stop all action on the set. It also, however, refers to the fastest method of getting from one scene to another by instantly replacing one picture with another.

cutaway—A camera shot related to but not involving the main activity being taped. Used to cover jumps in the action at edit points. Also used to show reaction or to provide a visual break in a long scene.

cyclorama strip lights—A strip light mounted horizontally at the top or bottom of a cyclorama to light it in a smooth or uniform manner.

dc—Direct current. Electrical current that maintains a steady flow and does not alternate.

decibel—A subjective measure of sound volume or strength.

definition—The degree of detail or sharpness in a TV picture.

degauss—The process of demagnetizing tape for reuse.

depth of field—The distance between the nearest and farthest objects that can be shown in focus within the scene without camera adjustment. Generally, the higher the f-stop (q.v.) setting, the greater the depth of field.

dew—VCR indicator that signals excessive amounts of moisture in the environment and shuts off the VCR.

Dichroic Daylight Conversion Filter—A lens filter that balances the color values of objects in direct sunlight so that they will match the values of scenes taped in artificial light.

Dichroic Mirrors—Special mirrors that reflect special wavelengths of the light spectrum while allowing others to pass through their semitransparent surfaces. Used in color TV cameras to divide light into the three primary colors for pickup tubes.

Die-cast Chassis—A VCR construction design that uses a single-piece annealed aluminum chassis on which are mounted the crucial VCR components.

Differential gain—In color TV, the change in gain, expressed in dB, for the 3.58 MHz color subcarrier as the level of the luminance signal is varied from blanking to white.

differential phase—Variation in phase of the color subcarrier of a TV signal as the level of the luminance signal is varied from blanking to white.

diffused light—Light that illuminates a relatively large area with an indistinct light beam. Diffused light, created by floodlights, produces soft shadows.

diffusion screen—A wire mesh placed in front of a light to reduce and soften the intensity of light.

dimmer—A device used for controlling the amount of light radiated from a luminaire. Common types are resistance, autotransformer, magnetic amplifier, silicon controlled rectifier or semiconductor, thyratron, and iris type dimmers.

direct drive—A motor drive mounted inside a cylinder that eliminates the usual rubber belt drives. This system produces a more stable picture.

directional light—Light that illuminates a relatively small area with a distinct light beam. Directional light, produced by spotlights, creates harsh, clearly defined shadows.

director—The person in the television production who has charge of all matters of action and composition in the production and who supervises in the efforts of technicians, cameramen, etc.

dissolve—A slow crossfade.

distant signal—TV signals that originate at a point too far away to be picked up by ordinary home reception equipment. Signals defined by the FCC as outside a broadcaster's license area. Cable systems are limited by FCC rules in the number of distant signals they can offer subscribers.

distortion—A condition during transmission or amplification in which the received signal waveform departs from that of the original transmitted waveform.

distribution plant—The hardware of a cable system. Amplifiers, trunk cable, and feeder lines attached to utility poles or fed through underground conduits like telephone and electric wires.

dolly—The wheels on the feet of a tripod. The action of moving the camera tripod.

down & under—A cue to the audio personnel to reduce the sound level under dialogue or to fade that sound out altogether.

downstream—Signals traveling from the headend to subscriber's homes.

dramatic—A video production of a theatrical nature.

dropout—A loss of picture signal during tape playback (displays as a black or white streak in the monitor).

dropout compensator—An electronic circuit that replaces a missing line of video information.

dual cable—A method of doubling channel capacity by using two cables installed side by side to carry different signals.

dubbing—The duplication of a videotape or the addition of new audio information to the tape.

dynamic microphone—A pressure-sensitive microphone.

earth stations—Ground terminals that use antennas and associated electronic equipment to transmit, receive, and process communications via satellite. Future cable systems may be able to interconnect by domestic communications satellites, creating regional and national cable networks.

easement—The right to use land for a specific purpose, such as laying cable.

ECU—Extreme close-up.

edit code—Electronic impulses recorded on the cue track or during the vertical interval to address each frame.

editing—The process of putting into a predetermined sequence the various segments of master tapes.

editor-programmer—Electronic control device that accurately and automatically programs some of all VTR functions during editing.

edit point—Audio or video event where an edit is to be made.

edit preview—An editing feature found on sophisticated automatic editing systems that allows the operator to preview his edit before it is actually made.

edit rehearsal—The second generation edited tape that has become the "new" master containing all the edited scenes in the proper sequence.

edit VCR—Videotape recorder with editing capability used to record material in sequence from master tapes or other sources.

EFP—Electronic field production.

EIAJ Type #1 Recommended Color Standard—The compatible color standard established by the Electronic Industries Association of Japan. Compatible with the EIAJ Type #1 Standard in that color tapes can be played back in black and white on EIAJ Type #1 black-and-white VTRs, and black-and-white tapes can be played back in black and white on EIAJ Type #1 color VTRs.

EJ—Electronic journalism.

electret condenser microphone—A very sensitive microphone that requires a power supply in the microphone and needs a battery.

electromagnetic spectrum—A continuous range of frequencies of electromagnetic radiation. Within the spectrum, waves have some specified common characteristics; the TV broadcast spectrum, for instance, ranges from 54 to 890 MHz.

electronic editing—The process whereby a videotape is edited.

electrostatic focus—A newer type of camera tube that is able to focus the electron beam without using heavy deflection coils.

ellipsoidal spotlight—A luminaire embodying a lamp, a reflector, a framing device, and a single or compound lens system, together with provisions for accommodating a pattern hold and patterns.

ENG—Electronic newsgathering.

erase head—A magnetic head that erases electronic information on the tape.

erasure—The neutralization of the recorded magnetic pattern on videotape. It is accomplished by placing the tape in a strong magnetic field. The process is called degaussing.

essential area—That area of a graphic which the camera will see.

establishing shot—Initial camera scene. Usually a long, wide shot that allows the viewer to orient himself to the location of the program.

ETOE—Electronics to electronics; a method of analyzing the output of a VTR relative to its input.

ETV—Educated television.

extended play tapes—Tapes that are longer than usual.

external microphone—A separate microphone that is not built into the camera.

eyelight—Illumination to produce a specular reflection from the eyes (and teeth) without adding a significant increase of light on the subject.

facsimile—The electronic transmission of pictures, charts, graphs, etc., from one place to another by radio, telegraphy or telephone. With special facsimile equipment, a home television receiver can deliver mail and newspapers by cable.

fade—A gradual change in the picture or sound intensity.

fader—A sliding pot control with which a signal is faded.

feedback—A loud squeal emitted from a speaker when a microphone is placed in the wrong relationship to it. Normally in front of the speaker.

feeder line—Intermediate cable distribution lines that connect the main trunk line to the smaller house drops that lead into residences.

ferrite—A brittle, high-frequency response material that is used to make video heads.

field—One-half of a complete TV scanning cycle, or 261½ scanning lines ; one-half of a 525-line TV frame.

fill light—Supplementary illumination to reduce shadow or contrast range.

film chain—An optical system whereby an image from a film or slide projector is transferred to a video camera for use in a television system.

filter—A glass element used to exclude certain types of light. Different filters must be used for differing light conditions.

first generation—The original recording of master tape.

flagging—Skew induced distortion produced at the top of the playback picture.

flat—The backdrops used on a TV studio set. They are made of hardboard, plastic, paper, and many other materials that are chosen depending on the need for one-time or multiple use.

flip stand—A system in which cards containing titles, artwork, or photographs are mounted in a ring binder. The material can be shown in sequence by flipping the cards forward (in front of the camera).

flood—Lights used for overall illumination. Also called scoops.

floor manager—A person who works on the studio floor during production and follows the director's orders with regard to the direction of on-camera talent. He also prepares the set for production efforts.

fluid-head tripod—A tripod head that moves very smoothly.

flying erase head—Erases existing information during the editing process. It is physically similar to the recording head. Leading the recording head, it erases frame by frame.

focal length—The distance between the center of a lens and where the image is in focus. Usually the image plane or the camera tube's target plate.

focus—The greatest possible resolution of an object when the object seems to be sharp and well defined.

follow focus—The continual adjustment of the lens to keep an object in focus while the object and/or the camera moves.

foot-candle—Generally measured in lumens per square foot. The amount of illumination from 1 international candle falling on a 1-square-foot surface at a distance of 1 foot.

format—A limited grouping of similar VTRs such as a ½-inch and ¾-inch or small format and large format.

frame—One complete TV picture.

frame rate—The speed at which the frames are scanned (30 frames a second for video).

franchise—Contractual agreement between a CATV operator and the governing municipal authority. Under federal regulation a franchise, certificate, contract or any other agreement amounts to a license to operate.

frequency—The number of times an electromagnetic signal repeats an identical cycle in a unit of time, usually one second. One hertz (Hz) is 1 cycle per second. A kHz (kilohertz) is 1000 cycles per second; a MHz (megahertz) is 1 million cycles per second; a GHz (gigahertz) is 1 billion cycles per second.

frequency response—The capability of an electronic device to reproduce a wide range of frequencies.

fresnel lens—A lens that acts similar to a plano-convex lens but is thinner and lighter due to steps on the convex side. Often the flat side has a rough surface to smooth light beams by slightly diffusing the light.

fresnel spotlight—A luminaire embodying a lamp and a fresnel lens, with or without a reflector, that has a soft beam edge. The field and beam angles can be varied by changing the spacing between the lamp and lens.

f-stop—A calibrated control to limit the amount of light entering the camera. The higher the f-stop number the more light is stopped. Therefore, dimly lit scenes require a smaller f-stop setting to allow more light into the camera. The f-stop setting also reflects the depth of field.

F-type connector—A metal screwlike connector used for connecting a VCR to an antenna or CTV system and to a TV set.

gain—The level of signal amplification.

gel cell—A lead gelled acid battery.

generation—The number of times a master tape has been duplicated. Successive generations introduce successively increasing amounts of noise into the signal. The number of generations becomes a factor in editing and duplicating decisions.

genlock—The locking up of the sync generators of one or several sources to a main sync source, i.e., a VTR locked up to an SEG.

glitch—Colloquial term for picture breakup due to loss of control track from dropout or mistimed edit.

GOBO—A two-dimensional piece of artwork placed between the camera and a set. Used to create a graphic illusion.

grandfathering—Exempting cable systems from the federal rules because they were in existence or operation before the rules or substantial investments were made in system construction before the rules. Grandfathering applies to signal carriage, access channels, and certification process.

gray scale—The various shades of gray in a TV picture that correspond to color.

guardbands—The separations between the video tracks on the tape.

halo—A condition caused usually by a bright object on a dark background. It causes a flare of light.

hardware—The equipment involved in production, storage, distribution or reception of electronic signals. In CATV it means the headend, the coaxial cable network, amplifiers, the television receiver and production equipment like cameras and video tape recorders.

harmonic distortion—Form of interference involving the generation of harmonics according to the frequency relationship $f = nf_1$ for each frequency present, where n is a whole number equal to 2 or more.

head—An electromagnetic device that records or retrieves information from magnetic tape.

head channel—The signal path unique to each magnetic head.

head clogging—The accumulation of debris (usually oxide particles) on the head—the usual result of which is a loss of signal during playback and degradation or failure to record during the record mode.

headend—Electronic control center—generally located at the antenna site of a CATV system—usually including antennas, preamplifiers, frequency converters, demodulators, modulators and other related equipment which amplify, filter and convert incoming broadcast TV signals to cable system channels.

head room—The space between the top of the performer's head and the top of the camera picture. The cameramen must allow some extra space so that the top of the performer's head is not cropped during dubbing.

helical scale—A helixlike method of recording video information on tape in which the signal is recorded diagonally in adjacent strips. Also called slant track recording.

hertz (Hz)—Cycles per second.

high density—Videotape that packs many magnetic particles per square inch on the tape.

high impedance—(HI-Z) 800-10,000 ohms. See impedance.

highlights—The maximum brightness of the picture or the areas that have the greatest illumination.

high VHF band—Part of the frequency band the FCC allocates to VHF broadcasting, including channels 7 through 13, or 174 through 216 MHz.

horizontal resolution—The number of vertical lines that can be perceived by a video camera in a horizontal direction on a test chart.

hot—A condition in which too much light is falling on a performer, an object, or a portion of the set.

house drop—The coaxial cable that connects each building or home to the nearest feeder line of the cable network.

HPF—Hot-pressed ferrite.

hue—A term used to describe the position of a color in a range that runs from red to yellow, to green, to blue, to violet, and back to red. All colors have a hue. Black, gray, and white do not have hue.

IC—Integrated circuit.

idiot cards—Cue cards.

image enhancer—Any electronic device that sharpens the picture.

image plane—The point at which the image is focused (generally at the camera tube target plate).

impedance—The resistance of a component to the flow of electrons—rated in ohms. Generally expressed as high impedance (hi/Z) or low impedance (low-Z).

in-camera editing—The technique of editing as you shoot.

incandescent light—The light produced by the hot filament of ordinary glass-globe light bulbs.

incident light—Light that strikes the object directly from its source. Incident light reading is the measure of light (in foot-candles) from the object to the light source. The foot-candle meter is pointed directly into the light source.

in-line color—A color TV tube system in which the three electron guns producing the primary colors of the color signal are next to each other in a straight line rather than in a triad, as has traditionally been the case in color TV manufacture.

insert edit—Adding new information to an existing video tape.

interchangeability—The capability to exchange tapes between different manufacturers' video tape recorders with no appreciable degradation of playback image.

interconnect—To link CATV headends, usually with microwave, so that subscribers to different cable systems can see the same programming simultaneously.

interference—Energy that tends to interfere with the reception of the desired signals (such as fading from airline flights, rf interference from adjacent channels, or "ghosting" from reflecting objects such as mountains and buildings).

interlace—The electronic process of scanning alternate lines (fields) of a TV picture so that flicker is minimized.

intermodulation distortion—Form of interface involving the generation of interfering beats between two or more carriers according to the frequency relationship $f = nf_1 + mf_2$, where n and m are whole numbers (but not zero), with appropriate expansion for additional carriers.

IPS—Inches per second—tape speed or writing speed.

iris—The adjustable set of leaves over the aperture of a lens. The part which actually moves when an f-stop setting is changed.

jack—A cable connector, as in phone jack, minijack, cannon jack.

joy stick—A lever that controls the position of a special effect on a screen or the motion of tape on a VTR during the editing process.

JVC—Japanese Victor Corporation.

Kelvin—A temperature scale where each degree is the same size as a degree centigrade but has its zero at −273°C; i.e. −273° −0°K, 0°C =

273°K, 100°C = 373°K, etc. This is the unit of temperature used to designate the color temperature of a light source.

keying—The matting of one picture over another.

key light—The principal source of light that establishes the character of the actor together with the atmosphere and mood of the scene.

kicker—A side or back light often near lens height. Used to rim faces and model profile shots.

kilohertz (KHz)—Thousand cycles.

kinescope recording—A film recording made by a motion picture camera designed to photograph a television program directly from the front of a television tube. Often called a "kine."

lavalier—A microphone worn on a lanyard around the neck. Lavaliers are larger and more obtrusive than minimicrophones but are less obtrusive than a hand held microphone.

lead-in—The opening portion of a program usually with voice-over narration or music.

leapfrogging—CATV operator's practice of skipping over one or more of the nearest TV stations to bring in a farther signal for more program diversity on the cable.

leaseback—The practice by telephone companies of installing and maintaining CATV distribution systems and "leasing" the facilities "back" to separate contractors for operation of the system.

leased channel—In addition to the designed channel and over-the-air broadcast channels, cable systems must make the remainder of their channels available for lease. At least one channel is to be available for lease by part-time users. Channels can be leased by group or individuals for the sole use of the lessee at posted rates. The operator must not censor or control program content in any way (under FCC rules).

lens—A series of optical elements, contained within a tube, that collect and focus light.

lens paper—A paper specially made for cleaning lenses.

lens speed—The measurement of the ability of a lens to collect light, usually expressed by its lowest f-stop number.

level—The average intensity of an audio or video source.

lighting grid—A fixed structure of either aluminum or steel members, such as pipe, located above the studio floor for the purpose of supporting luminaires and to support the electrical outlets required.

lighting triangle—Same as photographic principle. The triangular arrangement of key lights, back lights, and fill lights.

light lag—The tendency of a camera tube to retain its image causing a ghost-like effect.

light level—The intensity of the light as measured in foot-candles.

light ratio—The relative intensities of key, back, and fill. A 1:1 ratio between key and back lights means that both light sources burn with equal intensities. A 1:½ ratio between key and fill lights means that the fill light burns with half the intensity of the key light. Because light ratios depend on many other production variables, they cannot be fixed. A key: back:fill ratio of 1:1:½ is often used for normal triangle lighting.

limbo—A black background having no detectable detail.

line in—An audio input for other VTRs or an audio mixer.

line level—600 ohms; a low-impedance signal.

line out—Audio output.

lip sync—Perfect or "real" synchronization between sound and picture; a subject's lips move perfectly with his voice on the sound track.

live—The process of transmitting a program the instant it is taking place.

local origination channel—A channel on a cable system that is programmed by the cable operator and subject to his exclusive control.

local signals—Television signals that are received at locations within the station's predicted grade B contour. Cable systems must carry all commercial local signals in their areas.

lock out logic—An electronic circuit that prevents the improper sequencing of controls.

long lens—A telephoto lens.

long shot—A shot from a great distance.

low impedance—30 to 600 ohms. *See* impedance.

low light—Subdued illumination.

LP—Long play.

luminance—The photometric brightness or radiance of a light source.

luminance signal—That portion of the NTSC color TV signal which contains the luminance or brightness information.

lux—An old-fashioned measurement of light. One Ft–c=10.76 lux.

macro lens—A lens that is capable of close-up focusing.

magnetic tape—A tape consisting of a flexible-base material usually coated on one side with a thin magnetizable layer.

mask—Material used to cover some part of the set or picture that the director does not want included.

master tape footage—The original tapes.

match edit—Motion shot by two independent cameras, or one camera at two separate times, edited together to create the illusion of continuous motion.

matching transformer—An electromagnetic device that equalizes the impedance of two different input sources.

medium shot—The camera view between a close-up and a long shot.

megahertz (MHz)—Million hertz.

microwave—Line-of-sight, point-to-point transmission of signals at high frequency. Many CATV systems receive some television signals from a distant antenna location with the antenna and the system connected by microwave relay.

mid-band—The part of the frequency band that lies between television channels 6 and 7, reserved by the FCC for air, maritime and land mobile units, FM radio and aeronautical and maritime navigation. Mid-band frequencies, 108 to 174 MHz, can also be used to provide additional channels on cable television systems.

miniplug—A small audio plug. Sometimes called a Sony plug.

mistracking—The picture distortion caused by improper tape-to-head or tape-path contact.

mixer—A device that combines and blends two or more audio or video sources.

M-load—A type of cassette tape loading mechanism used in VHS VCRs.

modulator—The electronic equipment required to combine video and audio signals from a studio and convert them to radio frequencies (rf) for distribution on a cable system. Also, a very low-powered television signal generator used to provide signals for distribution on a CATV system.

moire—Herringbone interference patterns in a TV picture.

monitor—A television set that accepts direct cable feeds from TV cameras or broadcast signals. To monitor means to observe picture shading and other factors.

monitor/receiver—A dual-function standard TV receiver and monitor.

monochrome—Black and white.

movieola—Mechanical device used to edit film.

MSO (Multiple System Operator)—A company that owns more than one CATV system.

multiplexer—An optical device for using a single television camera in conjunction with one or more motion picture or slide projectors.

narrator—The performer, usually off-camera, who delivers lines of copy in a televised program. If he is entirely off-camera, then his words are designated as "voice-over."

negative image—A condition in which the polarity of the picture is exactly opposite that of normal, and the resulting picture appears black in white areas and vice versa.

net weekly circulation—The estimated number of television households viewing a particular station at least once per week, Monday through Sunday, 6 A.M. to 2 A.M. EST. Used to determine whether a station is "significantly viewed" in an area and must be carried by a cable system operating in that area.

network—Group of radio or television outlets linked by cable or microwave that transmit identical program simultaneously, or the company that produces programs for them. The three national commercial television networks in the U.S. are the American Broadcasting Company, the Columbia Broadcasting Company, and the National Broadcasting Company.

newvicon tube—A Panasonic camera tube that has very good sharpness and low-light characteristics.

Ni-Cad—Nickel-cadmium battery.

noise—Extraneous random signals that interfere with the desired signal. Video noise appears as random dots or "snow" on the screen. Noise increases with successive generations.

nonlens spotlight—A luminaire embodying only a lamp and a reflector that has variable field and beam angles obtained by changing the spacing between the lamp and the reflector.

NTSC—National Television Systems Committee. A committee that worked with the Federal Communications Commission in formulating standards for the present-day United States color television system.

NTSC color—The color standard used in the United States and set by the National Television System Committee. Compatible color that can be received in black and white.

oersteads of coercivity—The measurement of the ability of videotape to store information.

off the air—Reception of a TV signal that has been broadcast through the air.

omnidirectional—A microphone pickup pattern that is sensitive to sounds coming from all directions.

omnivision—Panasonic ½-inch VCR.

one-tube color camera—A color-capable video camera that produces a color signal through the use of only one pickup tube.

open-reel—A recording device that uses two exposed reels.

open-up—To set the f-stop to a lower number (allowing more light into the camera).

ordinance—A local law that governs the institution and operation of a cable television system in a community.

oscilloscope—An electronic device that measures electronic signals. It is used for alignment of video equipment.

overlapping markets—Where a TV station's predicted grade B contour is overlapped by the predicted B contour of a television station located in a different market.

oxide—The magnetic particles that are bonded on the tape.

pairing—Problems in which the interlace of scanning lines of alternate fields fall (in pairs) on top of each other instead of between each other.

PAL—Phase alternation line; Britist-German color TV standard.

Pan—A movement of the camera lens to the left or right.

pantograph—A hanger-type assembly having a scissor-type mechanical action that permits variable height adjustment and which is counterbalanced with tension springs and friction devices.

park point—Preroll point where VCR stops following a backspace.

patch panel—A plug and jack assembly that permits studio outlets to be temporarily connected to various dimmer and nondim circuit outputs.

pause—A VCR control that momentarily stops the tape.

pay TV—A system of television in which scrambled signals are distributed and are unscrambled at the homeowner's set with a decoder upon payment of a fee.

peak white—The brightest, whitest level which a picture signal can transmit.

pedestal—The minimum level which the blackest area of the video signals are allowed to reach.

penetration, CATV—The ratio of the number of subscribers to the total number of households passed by the cable system. Penetration is the basis of a system's profitability.

performance standards—The minimum technical criteria, that must be met by CATV systems, consistent with standards set by the FCC or the local ordinance.

phase—The relative timing of one signal to another.

phone plug—An audio plug with ¼-inch-wide connector shaft.

phono plug—RCA audio plug.

photographic principle—The triangular arrangement of keylights, back lights, and fill lights with the back light opposite the camera and directly behind the object, and the key and fill lights opposite each other to the front and side of the object.

pickup pattern—A determination of the direction(s) from which a microphone is sensitive to sound waves.

picture tube—The cathode-ray tube of a television monitor or set. It creates an image by variations of an electron beam that scans the coated surface of the interior of the vacuum tube.

plumbicon—The trade name of Philips' special vidicon tube that is more sensitive than a normal vidicon; used in some color cameras.

pole attachment—When CATV systems use existing pole lines maintained by power and telephone companies, an attachment contract must be negotiated between the parties of interest.

pole tips—Those parts of the video head that protrude radially beyond the rim of the head wheel and form the magnetic path to and from the tape.

portapak—A portable video system.

post-production—Assembly of various program elements by editing in the studio following original shooting.

pot—Audio term for a volume control knob; short for potentiometer.

preamplifier—An amplifier to raise signal levels from microphones, etc., to a point high enough to be usable by a power amplifier.

predicted grade A contour—The line representing the service area in which a good picture is computed to be available 90 percent of the time at 70 percent of the receiver locations. Signal contours determine what educational channels are carried on a cable system and, in smaller markets, what stations must be carried from other small markets.

predicted grade B contour—The concentric area marking a television station's service area in which a good picture is computed to be available

90 percent of the time at 50 percent of the receiver locations.

preproduction—Preparation activity that occurs prior to actual shooting.

preroll—The process of backing up tapes and VTRs in preparation for an edit.

primary colors—The three colors used in color TV. No two of them can be combined to produce the third. Red, blue, and green.

print-through—A video duplication process involving the transfer of the signals of a master tape to a slave tape by magnetic induction.

program selector—A control of VCRs which selects the VCR output to the TV set.

public access—To ensure that divergent community opinion is aired on cable television. The public access channel is free and available at all times on a first-come, first-served basis for noncommercial use by the general public. CATV systems may offer one or more public access channels.

public TV—Noncommercial television broadcasting.

pulse-cross monitor—Monitor display device allowing the monitoring of tension and tracking of VTRs during playback.

quadruplex—A video recording system using four video heads to scan the tape at a 90-degree angle.

quartz lighting—Very bright and efficient lighting.

raster—The scanned or illuminated area of a television picture tube.

RBG signal—The chrominance information; red, blue, and green.

rear-screen projection—A process in which a picture (either slide or motion picture film) is projected from the back of a translucent screen. Performers can then work in front of this "background" as though it were real.

receiver isolation—The attenuation between any two receivers connected to the system.

reel—A hub with flanges on which videotape or film is wound.

reference black level—The picture signal level corresponding to a specified maximum limit for black peaks.

reference white level—The picture signal level corresponding to a specified maximum limit for white peaks.

remote—Shooting tape or film outside the studio. Also called "on location."

resolution—The amount of resolvable detail in a picture.

retrofocus—A lens with close-up focusing capability.

rf—Radio frequency. The range of frequencies used to transmit electric waves.

rf amplifier—An amplifier used for amplifying rf signals.

rf converter—A device that allows video and audio signals from a VTR to play back on a standard TV set.

roll—Starting of VTR play or record.

RG-59U—Coaxial video cable.
rotary erase head—*See* flying erase head.
RPM—Revolutions per minute.
running time—The estimated (and later the actual) time it takes for the presentation of a given show.

saturation—The vividness of a color described by such terms as bright, deep, pastel, pale, etc. Saturation is directly related to the amplitude of the chrominance signal.
scanning—The horizontal and vertical movement of the electron beam in the camera or TV tube.
scene—A sequence in a television program. It can consist of one or more TV shots. The term may also refer to the setting itself; the "scene" in which the action is occurring.
scoop—A luminaire consisting of only a lamp and reflector with fixed spacing. Generally, the reflector has a diffused finish and is large in size.
scrim—A flag made of translucent material, such as gauze or glass fabric, and used to diffuse or soften light. The term is also applied to gauzelike theatrical curtains. When properly lighted from the front, this curtain can appear opaque. If light is brought up behind the scrim, it becomes transparent.
search—A VCR function that causes it to locate or stop at a particular point on a tape.
SECAM—Sequential Couleur a Memoire. A color TV system developed by the French that differs radically from both the NTSX and PAL color systems.
set light—Separate illumination of background or set other than that provided for principal subjects or areas.
shadow mask color tube—RCA-developed standard color tube; a color tube equipped with a metal sheet with half a million small holes punched in it. The metal sheet (which is the shadow mask) is placed between the electron guns, which beam the picture signal, and the phospher-coated screen.
shooting ratio—The amount of tape recorded relative to the amount of tape actually used.
shooting script—A TV script that includes all copy and related camera shots, sound effects, lighting plans, music, and other pertinent production information.
shot—The pickup from a television camera.
shotgun microphone—A highly sensitive microphone that allows good sound pickup from a great distance.
signal—Information transposed into electrical impulses.
signal level—The rms voltage measured during the rf signal peak. It is usually expressed in microvolts referred to an impedance of 75 ohms, or in dBmV, the value in decibels with respect to a reference level of 0 dBmV, which is 1 millivolt across 75 ohms.

signal-to-noise ratio—The amount of visible "snow" or "noise" is a given picture relative to the actual picture signal. The higher the signal-to-noise ratio the better the picture.

single system—Film sound recording where sound is recorded directly onto the original film as it is exposed; c.f. "double system" where sound is recorded separately and later matched up or put in lip sync. Video recording is by nature single system.

skew—The tape tension.

skip-field recording—A recording process whereby only one field is reproduced twice instead of reproducing both fields and interlacing them.

small format—The small-width video tape equipment such as ¼ inch, ½ inch, ¾ inch and 1 inch.

smear—A picture condition in which objects appear to extend horizontally beyond their normal boundaries or edges. A blurred or "smeared" effect is the result.

sneak—A gradual introduction of a sound, a dissolve, music, etc., so that its presence isn't immediately obvious to the audience.

sock cue—To introduce music very suddenly at peak volume. It is also called a "stinger" and can be used for emphasis and shock value.

soft light—A well-diffused light source that produces soft or poorly defined shadows when an object is placed in between the luminaire and the light background.

software—Programming and programming materials such as films, videotapes, and slides.

solenoid—An electromagnetic circuit control system.

solid-state—Electronic circuitry that contains no vacuum tubes.

source VTR—In editing with two VTR's, the source VTR plays back the original tape which is then recorded onto the editor VTR.

SP—Standard play.

special effects—Fades, dissolves, wipes, split screens, keys, chromakeys, mattes, etc. In television, special effects are inserted directly between two sources using a special effects generator (SEG). In film, special effects are sometimes inserted "in the camera" by multiple exposures (titles) or more commonly by use of an optical printer.

special effects generator—A device permitting combinations of images on a television screen supplied by one or more video inputs.

speed switch—A VCR control that selects one of two different speeds.

spindle—The shaft on a tape recorder that passes through the drive hole in the tape reel.

splice—A butt joint between two pieces of tape (or film) held together by means of a strip of self-adhesive.

splicing tape—A special material used to splice magnetic tape.

split screen—An electronic effect in which two different pictures appear side by side on the same screen. It can be accomplished with the use of a special-effects generator.

splitter—A passive device (one with no active electronic components) that distributes a television signal carried on a cable in two or more paths and sends it to a number of receivers simultaneously.

spotlight—A light instrument that produces directional, relatively undiffused light.

spread—To enlarge (diffuse) the light beam of a spotlight by pushing the light-bulb reflector unit toward the lens or the lens closer to the light bulb.

stand by—The cue that indicates to performers and crew that taping is about to begin. It is preceded by the request of "Quiet in the studio, please."

stop down—To adjust the f-stop of a lens to a higher number, allowing less light to enter the camera.

straight dubbing—The process of simply copying a tape in real time without editing.

strike—To dismantle a set or take away props.

stripe filter—A chrominance tube system in which the target area of the tube is divided into sequential striped for R, B, G, and Y, and can therefore derive a color signal by using only one pickup tube.

striplight—A luminaire with a number of lamps arranged in a line. Often each lamp is in an individual compartment. There could be a reflector behind the lamp or a color media in front of the lamp. Striplights are normally wired in three or four circuits.

subband—The frequency band from 6 MHz to 54 MHz. It may be used for two-way data transmission.

subcarrier frequency—The frequency on which color information is modulated in the color TV system; in the United States it's 3.58 MHz.

subcarrier phase shifter—Special circuitry designed to control the phase relationships of the two portions of the encoded color signal so that they maintain their correct relationship during recording, transmission, and reproduction. A phase shifter allows the user to change the timing of the signals involved so that they are occurring at the correct time and are said to be in phase.

subscriber—A person who pays a fee for cable services.

super band—The frequency band, from 216 to 600 MHz, used for fixed and mobile radios and additional television channels on a cable system.

superimpose—To place the image from one source over the image from another source. It is called a "super."

switched service—Cable communications service in which each subscriber has a terminal and may communicate with any other subscriber.

switcher—A control that permits the selection of one image from any of several cameras to be fed into the television display or recording system.

switcher/fader—A device permitting gradual, overlapping transition from the image of one camera to another. Sometimes incorporated as part of a special effects generator.

sync—An abbreviation for synchronization. It refers to the process of producing synchronizing signals, or timing pulses, that keep various television circuits in phase with one another.

sync generator—A device used to supply a common or master sync signal to a system of several cameras. This ensures that scanning pulses will be in phase. Scanning pulses out of phase produce distortion or rolling, sometimes called sync loss.

synchronization—Maintenance of simultaneous audio and video signals so that sound and picture operate in phase. Abbreviated "sync."

system noise—Refers to the random energy generated by thermal and shot effects in the system. It is specified in terms of its rms level as measured in a 4 MHz bandwidth centered within a 6 MHz cable television channel.

talent—The performers who appear in a television production.

tally light—The red light on a TV camera, when lit, that indicates it is the taking camera.

tap—A device installed in the feeder cable that connects the home TV set to the cable network. Also called a drop.

tape configuration—Tape packaging.

tape guides—Grooved pins on either side of the recording head assembly that serve to position the magnetic tape so that it feeds properly into the head in either the record or playback mode.

tape path—The path of the videotape past the heads.

target—The image forming area of the camera area.

tariff—A schedule of services and rates filed by common carriers with regulatory bodies.

tearing—A picture condition when the image is displayed horizontally; usually caused by horizontal sync problems.

technical director—The person in charge of the electronic portion of the recording session. He might also do switching for the director.

television households—A television household is a household having one or more television sets. Estimates for each country are based on an updating of ARB (American Research Bureau) TV penetration estimates. The number of households that watch television in any locality during prime-time hours determines the top 100 markets.

television market—A city or complex of closely associated cities served by commercial television broadcast signals from one or more TV stations located within the area. The FCC uses television markets for designating what kind of cable services an operator should provide in terms of signal carriage and nonbroadcast channel use.

terminal—The connectors, transformers, and converter (if necessary) on the cable subscriber's set.

termination—The resistance value placed across the receiving end of a coaxial cable (video signals are terminated by 75 ohms).

test pattern—A specially made pattern of lines or circles that is used for setting focus, linearity, registration, balancing, and otherwise lining up a television camera.

third generation—Two copies removed from the original tape.

three-tube color camera—A color-capable camera that produces a color signal through the use of three pickup tubes, each assigned to one of the primary colors. An early stage in the development of the color video camera (introduced by RCA in 1940).

throw—To direct the light of an instrument in a particular direction. Also the effective distance between a luminaire and the area being lighted.

tight edit—An edit that has the minimum timing disturbance at the point of edit.

tight shot—Close-up.

tilt—To swivel the camera on its tripod. Either "tilt up" toward the ceiling or "tilt down" toward the floor.

time base corrector—A device that electronically corrects the mechanical and electronic errors created by VTR.

time base stability—Variation of playback speed to original recording speed.

top 100 market—Ranking of largest television broadcast areas by size of market, i.e., number of viewers and TV households. Used in selling air time to advertisers.

tracking—A playback adjustment on VTRs to ensure that the scanner is tracking properly.

tracks—The area on a tape that contains the video or audio information.

transfer—A dub or a copy.

translator—A type of relay system that picks up signals from distant or blocked out television stations, converts the signals to another channel to avoid interference, and retransmits them into areas the original signals could not have reached. Translators do not use cable to reach subscribers' homes and do not offer the other kinds of services that cable can provide (local origination, two-way operation, etc.).

trinicon tube—A Sony color tube that contains 3 color filters.

tripod—The three-legged stand upon which a camera is mounted.

trucking—To move the camera and tripod. See dolly.

trunk line—The major distribution cable used in CATV. It divides into feeder lines that are tapped for service to subscribers.

turret—A camera mounting that has accommodations for up to four lenses. The mount rotates to bring successive lenses into position in front of the camera.

TV penetration—The percentage of homes having one or more television sets at the time of the ARB survey. The ARB surveys local markets from October through July; the number of surveys in a year depends on the size of the market.

two-way capacity—A CATV system with two-way capacity can conduct signals to the headend as well as away from it. Two-way bidirectional systems now can carry data as well as full video television signals in either direction.

UHF—Ultra high frequency; a connector used for video cables.
U-load—A type of cassette loading mechanism used in U-Matic VCRs.
U-Matic—Sony trademark name that refers to the ¾-inch video cassette format VTRs.
unidirectional—A selective microphone pickup pattern.
upstream—Signals traveling from subscribers to the headend.

variable focal length lens—*See* zoom lens.
V-cord II—Sanyo two-hour, ½-inch video cassette recorder.
VCR—Video cassette recorder.
VHF—Very high frequency.
video—The visual components of a television signal.
video cassette—A self-contained video tape module for use in specially designed video tape equipment.
video distribution amplifier—An amplifier that strengthens video signals.
video gain—The amplitude of the video signal.
video in—A video input.
video out—A video output.
Video Star—A VCR manufactured and distributed by JVC.
video switcher—*See* switcher.
videotape—The mylar strip, similar to audio recording tape, on which the video and audio signals are recorded. Unlike film, videotape can be instantly replayed, and it can be used over and over again.
video tape recorder (VTR)—An electro-mechanical device capable of recording and reproducing an electronic signal onto videotape. When malfunctioning, it is also capable of transforming the videotape into crochet material.
video transmission—The original video signal before it is modulated and converted to radio frequency and broadcast or cablecast. A home television set reconverts radio frequencies to a video signal.
vidicon tube—An inexpensive and versatile electronic pickup tube used in TV cameras.
viewfinder—A small black-and-white video monitor connected to a camera that the camera operator uses in composing the picture.
vignetting—An effect produced by a lens when it fails to completely cover the camera tube image area.
Visc—A videodisc device marketed by Matsushita.
voice over—A spoken message or narrative added to the picture.
VTR/VCR—Abbreviation for video tape recorder, video cassette recorder, a device which allows the recording and playback of magnetic tape sound and picture recordings.
VU meter—Volume unit meter. Measures audio levels.

watt—A unit of electrical power.

waveform monitor—An oscilloscope used to display video signal strength.

white balance—The video camera's capability to produce white correctly.

white clipper—A circuit designed to limit white peaks to a predetermined level.

white compression—Amplitude compression of the signals corresponding to the white regions of the picture. This results in differential gain.

white peak—The maximum excursion of the picture signal in the white direction.

white set—The adjustment of a color camera to correct white reference.

wide angle lens—A lens with a very wide angle of view.

wide open—Descriptive of a lens set at its lowest f-stop (q.v.) rating, letting in the maximum amount of light.

wipe—A special effect whereby one image "pushes" another image off the screen.

OW—Variations in tape speed that cause subsequent distortion of audio signals.

writing speed—The resultant difference between the speed of the video heads and the speed of the videotape.

X1—One-hour play/record mode on Beta VCRs.

X2—Two-hour play/record mode on Beta VCRs.

Y—The symbol for the luminance portion of a color video signal; the color video signal consists of R, G, B, and Y.

zoom—A multipurpose lens that is continuously variable everywhere from a wide angle to a telephoto position. To "zoom in" means to go for a shot in which the subject of interest slowly becomes larger in the picture. To "zoom out" means to pull back to a much wider shot.

zoom lens—A variable focal length lens that can change the apparent distance of the camera to the subject without moving the camera position.

Index

A

Amplifier, processing, 12
Animation, 64
Arc, 98
Artwork, 86
Audio advice, 20
Audio dub, 33
Audio grammar, 107
Audio production, planning your, 35
Audio recording checklist, 31
Audio sweetening, 107
Audio transitions, 107
Automatic gain control, 16

B

Back light, 44
Backspacers, 54
Barndoors, 42
Bedroom documentaries, 134
Beta, 4
Broadcast dub, 114
Burst, 18

C

Camera, viewfinder, 69
Camera, walking through the, 105
Camera angles, 99-101
Camera controls, 42
Camera movements, 96
Camera point of view, 98
Camera preview monitor, 69
Cameras, matching, 70
Cameras, tubeless, 3
Cardioid, 23
Character generator, 91
Chroma, 19
Color, deriving, 8
Color and television, 7
Color correction, 80
Colored backgrounds, 91
Color playback, 9
Color temperature, 39
Computer, home, 93
Condenser microphone, 24
Contrast ratio, 80
Convergence, 9
Crystal microphone, 24

D

Defocus, 105
Dimmers, 42
Directing the show, 76
Direct lighting, 44
Dissolver, 81
Dolly, 96
Double shoot, 59
Dub, broadcast, 114
Dubbing, 33

E

Editing, electronic, 52
Editing, live-action, 62
Editing control units, 54
Editing modes, 55

Editing on the run, 61
Editing setups, 55
Editing strategies, 58
Editing tips, 67
Electret condenser microphone, 24
Electronic editing, 52
Electrons, fluctuating stream of, 2
Electrons, steady stream of, 2
Environmental view, 133
Equipment and crew, 126
Eyeline, the, 101

F

Fades, 91
Field, 2
Fill light, 44
Film-to-tape transfer, 84
Frame, 2
Framing, 80
Frequency response, 21

G

Gels, 42
Genlock device, 90
Graphics, 126

H

Helical scan recording, 4
Hertz, 21
Home computer, 93
Horizontal blanking, 19

I

Impedance, 10, 24
Impedance mismatch, 10
Indirect lighting, 44
Instant replay, 115

J

Jumpcuts, 59

K

Key light, 44

L

Lecture and slide presentation, 81
Lens, close-up, 88
Lens filters, 90
Lighting, indirect, 48
Lighting, mood, 49
Lighting, setting up three-point, 46
Lighting, two-point, 48
Lighting methods, 39
Lighting the environment, 42
Light intensity, 39, 41
Light intensity, monitoring, 42
Line transformer, 26

M

Media, incorporating other, 79
Metal oxide semiconductor, 3
Microphone, condenser, 24
Microphone, crystal, 24
Microphone, dynamic, 24
Microphone, electret condenser, 24
Microphone, ribbon, 24
Microphone, selecting the, 21
Microphone, wireless, 24
Microphone mixers, 27
Mixers, microphone, 27
Mix mode, the, 34
Mix-up, 34
Monitoring the RF and Video signal, 15
Monitoring TV signals, 13
Monitor program, 69
Monitor waveform, 16

O

Omnidirectional, 21

P

Pan, 96
Pan, swish, 106
Pedestal, 18
Pick-up pattern, 21
Playback, color, 9
Playback, video, 6
Preproduction guidelines, 108
Preproduction information, 128
Presence, 23
Proc amps, 12
Production coordinator, 126
Production models, 114
Production precautions, 38
Production tips, 36, 49, 65, 78
Programs, fully scripted, 61
Programs, no-script, 58

R

Receiver/monitor, using a, 14
Recording, helical scan, 4
Refocus, 105
Replay, instant, 115
Reverberation, blanketing, 34
Reversals, 91
Role play, 119
Role play models, 120
Ruggedness, 23

S

Screen, hole in the, 82
Screen direction, 103
Scrims, 42

Script development, 125
Script directions, 77
Scripting in the field, 60
SEG, the, 69
SEG operation, 73
Sexual surveillance, 134
Signal, sync, 2
Signal monitoring, 13, 15
Signals, high-level, 27
Signals, stabilizing, 12
Signal splitting, 11
Slide-to-tape transfer, 79
Sound, 20
Sound, processing, 35
Sound effects, 107
Source gear, 55
Subject matter specialist, 125
Swish pan, 106
Sync, 18
Sync signal, 2

T

Tape, training, 124
Tapes, logging and indexing, 63
Television, color and, 7
Television picture, recording the, 3
Television picture, the, 1
Three-point lighting, 44
Tilt, 97
Time base corrector, 13
Titles, 86
Titles, insert, 91
Transformer, line, 26
Transitions, 103
Tripod, 87
Truck, 97
TV format, 79
TV process, the, 1
TV set monitoring, 13
TV signal, controlling and monitoring the, 10

U

Unidirectional, 22

V

VCR, 3
VCR fantasy skits, 135
VCR foreplay, 134
Vertical blanking, 19
VHS, 4
Video, body, 115
Video, day-in-the-life-of, 132
Video, legal, 129
Video, special event, 123
Video catalog, the, 122
Video deposition, 130
Video duplication, 65
Video expressions, 114
Video feedback, 91
Video fun and games, 114
Video grammar, 95
Video level, 16
Video playback, 6
Video scrapbook, the, 122
Video signal monitoring, 15
Video star, 129
Video will, the, 132
VTR, 3

W

Wattage, 42
Waveform monitor, 16
Wireless microphone, 24

Edited by Steven Bolt